Earth Observation for Integrated Water and Basin Management: New Possibilities and Challenges for Adaptation to a Changing Environment

Remote Sensing & Hydrology Symposium 2018

Proceedings of the International Association of Hydrological Sciences (PIAHS) Volume 380, 2018

Cordoba, Spain
8-10 May 2018

Editors:

M. P. Gonzalez-Dugo
A. Andreu
M. J. Polo
C. Neale
R. Pimentel

ISBN: 978-1-5108-8294-2

Printed from e-media with permission by:

Curran Associates, Inc.
57 Morehouse Lane
Red Hook, NY 12571

Some format issues inherent in the e-media version may also appear in this print version.

Printed by Curran Associates, Inc. (2018)

International Association of Hydrological Sciences
Centre for Ecology and Hydrology
Wallingford, Oxfordshire OX10 8BB
United Kingdom

Phone: 441 491 692 442
Fax: 441 491 692 448

http://iahs.info/

Additional copies of this publication are available from:

Curran Associates, Inc.
57 Morehouse Lane
Red Hook, NY 12571 USA
Phone: 845-758-0400
Fax: 845-758-2633
Email: curran@proceedings.com
Web: www.proceedings.com

TABLE OF CONTENTS

Proc. IAHS, 380, 1–2, 2018
https://doi.org/10.5194/piahs-380-1-2018

Open Access PIAHS

Preface: Earth Observation for Integrated Water and Basin Management: Challenges for adaptation to a changing environment

María J. Polo[1], Maria P. González-Dugo[2], and Christopher Neale[3]

[1]Andalusian Institute for Earth System Research, University of Cordoba, Córdoba, Spain
[2]Andalusian Institute of Agricultural and Fisheries Research and Training (IFAPA), Córdoba, Spain
[3]Robert B. Daugherty Water for Food Global Institute, University of Nebraska, Lincoln, USA

Correspondence: María J. Polo (mjpolo@uco.es)

Published: 18 December 2018

Integrated river basin management involves a sound knowledge of water and land interactions, and impacts from and feedbacks to human activity. Remote sensing has been an efficient and increasingly promising means of gathering direct information of the Earth surface, as well as information on water and energy fluxes. The recent generation of high-resolution sensors offers a huge potential for monitoring, assessing, and modelling our changing environment in a context of uncertainty about how future climate conditions will affect the current water resource and basin management framework. Moreover, large amounts of data are now available posing a challenging opportunity to the scientific community for both exploring and transforming these data into readily usable information products for different end-users in our societies.

The scientific decade 2013–2022 of IAHS, entitled "Panta Rhei – Everything Flows", is dedicated to research activities on change in hydrology and society. Undoubtedly, remote sensing from different sources provides us with the information needed to monitor the environment at different spatial and temporal scales. Renewed efforts are needed to merge different data sources for generating medium to long-term series that benefit from both earlier satellite missions and the recent availability of high quality data provided by the new satellites and airborne technologies.

The Remote Sensing and Hydrology Symposium (RSHS) is organized every four years by the International Commission of Remote Sensing (ICRS) of the International Association of Hydrological Sciences (IAHS). This conference is a meeting point for researchers from both the hydrology and remote sensing communities, and brings together their joint experience in Earth Observation for developing innovative and environmentally-sustainable water resources and basin management. The 4th edition, RSHS'18, was hosted by the Andalusian Institute for Earth System Research at the University of Cordoba, Spain, during 8–10 May 2018, who provided the venue facilities and sponsored the conference, with support from the IFAPA in Cordoba, and the Daugherty Water for Food Global Institute at the University of Nebraska.

RSHS'18 focused on the use of the new generation of remote sensors and applications to hydrology, water resources, agricultural water management and river basin management, and the challenges of blending remote sensing images and data sources with multi-scale modelling and ground-based data. Under the light of these objectives, innovative and integrated remote sensing research and applications were presented and discussed within the different thematic areas of the Symposium:

- Current and future missions for water cycle observation
- Observations of water cycle components
- Earth Observation retrievals and data products linked to the water cycle
- Applications of remote sensing data in water resources management
- Crop irrigation management by remote sensing
- Water quality and soil cover assessment from remote sensing data

This volumen gathers a selection of the works presented in RSHS'18. The Organizing Committee thanks all the people who made the conference a successful research discussion forum. IAHS provided sound institutional support and framework for the conference development. Three brilliant keynote lectures brought cutting-edge research on hydrological applications of remotely sensed data from different domains in the spectrum, and motivated questions for the discussion sessions:

- William Kustas, USDA-ARS, USA

  The Grape Remote sensing Atmospheric Profile and Evapotranspiration eXperiment (GRAPEX)-A sinopsis

- José Moreno, University of Valencia, Spain

  Vegetation fluorescence as a tool to monitor plant stress conditions: The FLEX Earth Explorer mission

- Claudia Notarnicola, EURAC, Italy

  Addressing hydrological challenges by exploiting remotely sensed imagery in the microwave domain

The work carried out by Elisabet Carpintero, Pedro Gómez-Giráldez, Elena Herrera, and María-José Pérez-Palazón in the Local Committee to produce an enjoyable time during the technical sessions and the social events is also acknowledged.

The work of the Scientific Committee is especially acknowledged, who revised the 127 abstracts submitted to the Symposium, and finally accepted 40 as posters and 87 as oral presentations:

- Martha Anderson, USDA-ARS Hydrology & Remote Sensing Lab
- Ana Andreu, University of Berkeley
- Gilles Boullet, IRD-CESBIO
- Alfonso Calera, University of Castilla-La Mancha
- Yangbo Chen, Sun Yat-sen University
- Michael Cosh, USDA-ARS Hydrology & Remote Sensing Lab
- Guido d'Urso, Universita' degli Studi di Napoli Federico II
- Simon Gascoin, CNRS-CESBIO
- Maria P. González-Dugo, IFAPA-Junta de Andalucía
- Christopher Hopkinson, University of Lethbridge
- William Kustas, USDA-ARS Hydrology & Remote Sensing Lab
- José Martínez Fernández, University of Salamanca
- Pamela Nagler, U.S. Geological Survey
- Christopher Neale, University of Nebraska
- Claudia Notarnicola, EURAC
- Rafael Pimentel, Swedish Meteorology and Hydrology Institute
- Maria J. Polo, University of Cordoba
- Andreas Schumann, University of Bochum
- Bob Su, University of Twente
- Julie Zinnert, Virginia Commonwealth University

56 works out from the oral presentations were selected by the Scientific Committee as full papers for the PIAHS volume of the conference. The Editors would like to thank Mario Ebel and Sarah Schneemann at Copernicus for their work and valuable support during the abstract submission process and the production of this volume.

Proc. IAHS, 380, 3–8, 2018
https://doi.org/10.5194/piahs-380-3-2018

Joint editorial: Invigorating hydrological research through journal publications

Nevil Quinn[1], Günter Blöschl[2], András Bárdossy[3], Attilio Castellarin[4], Martyn Clark[5], Christophe Cudennec[6], Demetris Koutsoyiannis[7], Upmanu Lall[8], Lubomir Lichner[9], Juraj Parajka[10], Christa D. Peters-Lidard[11], Graham Sander[12], Hubert Savenije[13], Keith Smettem[14], Harry Vereecken[15], Alberto Viglione[16], Patrick Willems[17], Andy Wood[18], Ross Woods[19], Chong-Yu Xu[20], and Erwin Zehe[21]

[1]Editor, *Hydrology Research*
[2]Past Editor, *Water Resources Research*; Editor, *Hydrology and Earth System Sciences*; Co-Editor, *Journal of Hydrology and Hydromechanics*
[3]Editor in Chief, *Journal of Hydrology*
[4]Co-Editor, *Hydrological Sciences Journal*; Associate Editor, *Water Resources Research*
[5]Editor in Chief, *Water Resources Research*
[6]Editor in Chief, *Proceedings of the International Association of Hydrological Sciences*
[7]Co-Editor, *Hydrological Sciences Journal*; Editor, *Hydrology and Earth System Sciences* (now retired)
[8]Editor in Chief, *Water Security*
[9]Editor in Chief, *Journal of Hydrology and Hydromechanics*
[10]Associate Editor, *Water Resources Research*; Co-Editor, *Journal of Hydrology and Hydromechanics*
[11]Chief Editor, *Journal of Hydrometeorology*
[12]Editor in Chief, *Advances in Water Resources*
[13]Editor in Chief, *Physics and Chemistry of the Earth*
[14]Editor in Chief, *Ecohydrology*
[15]Editor, *Vadose Zone Journal*
[16]Associate Editor, *Water Resources Research*; Associate Editor, *Hydrological Sciences Journal*
[17]Co-Editor in Chief, *Journal of Hydrology: Regional studies*
[18]Editor, *Journal of Hydrometeorology*
[19]Co-Editor, *Hydrological Sciences Journal*
[20]Editor, *Hydrology Research*
[21]Chief Executive Editor, *Hydrology and Earth System Sciences*

Correspondence: Nevil Quinn (nevil.quinn@uwe.ac.uk)

Received: 31 July 2018 – Published: 18 December 2018

Abstract. Editors of several journals in the field of hydrology met during the General Assembly of the European Geosciences Union (EGU) in Vienna in April 2017. This event was a follow-up of similar meetings held in 2013 and 2015. These meetings enable the group of editors to review the current status of the journals and the publication process, and to share thoughts on future strategies. Journals were represented at the 2017 meeting by their editors, as shown in the list of authors. The main points on invigorating hydrological research through journal publications are communicated in this joint editorial published in the above journals.

1 Introduction

Over the past five years, the editors of a number of journals in the discipline of hydrology have met informally to discuss challenges and concerns in relation to the rapidly changing publishing landscape. Two of the previous meetings, in Gothenburg in July 2013 and in Prague in June 2015, were followed by joint editorials (Blöschl et al., 2014; Koutsoyiannis et al., 2016) published in all participating journals. A meeting was convened in Vienna in April 2017 (during the General Assembly of the European Geosciences Union – EGU) which was attended by 21 editors representing 14 journals. Even though the journals are published in quite different settings, the editors found common cause in a vision of the editor's role beyond just that of gatekeeper ensuring high-quality publications, to also being critical facilitators of scientific advances. In that enabling spirit, we as editors acknowledge the need to anticipate and adapt to the changing publishing landscape. This editorial communicates our views on the implications for authors, readers, reviewers, institutional assessors and the community of editors, as discussed during the meeting, and subsequently.

2 Recent trends in the publication process – quantity, speed and multiple authorships

The previous joint editorials have reflected on the increased productivity across the discipline, and more broadly in science, as evidenced by a rise in manuscript submissions. This growth in submissions and publications has continued in recent years at an unfaltering rate. Collectively, the 14 journals represented in this editorial published 46 000 pages in 2017, compared with only 26 000 pages a decade earlier. The main driver of increased submissions has been intensified publication pressure, which has given rise to a number of trends of concern that privilege quantity over quality of science: in "salami publishing" (Martin, 2013; Koutsoyiannis et al., 2016) authors split a body of work into several papers in order to increase the number of their publications and their citation counts. There is also a tendency to publish work prematurely, where the contribution is incremental rather than significant. Despite the standard use of plagiarism detection tools by most journals, plagiarism still does occur, and "recycling", where authors repackage their own work with minimal extension for a different audience, is on the increase. Some of this would be regarded as self-plagiarism (Martin, 2013). There have been cases of authors submitting the same manuscript simultaneously to multiple journals, and authors immediately submitting a rejected manuscript to another journal without any reflection or revision in response to reviewer evaluations. There are also instances of reviewers (and editors) attempting to promote their own (or their journals') citation metrics by requiring authors to cite their list of papers (citation coercion and citation stacking). None of these practices are conducive to advancing the science of hydrology. On the contrary, they contribute to a system overload and a dilution of useful information in the published literature.

Another trend that has become acute recently is that of a push towards speedier publication. New media have created a culture of immediacy for traditional journals (Brossard and Scheufele, 2013), and editors are under pressure to reduce turn-around times, both in relation to time to first decision and the subsequent review process. Most hydrology journals have reduced their turn-around times by at least two months in the last decade, little of which can be attributed to technical and system improvements. A number of journals have introduced a "fast-track" or "rapid communication" route in an attempt to report quickly on an extreme event or new technology. These types of papers place a higher burden on reviewers in relation to speed and additional challenges to editorial teams regarding review quality, while authors risk compromising quality for expediency. Recent experience has highlighted the additional risks of premature press releases, where a paper is subsequently rejected but broadcasters have already acted on a press release. Various approaches to providing a "fast-track" stream are being considered by hydrology journals, with varying degrees of success. As a discipline we need to reflect on whether these approaches are consistent with the notion of high-quality communication in our journals or whether other communication forms (e.g. newsletters, professional magazines, new media) might be more appropriate. It may well be that different approaches may coexist within hydrology.

The third, conspicuous trend is that of an increase in the number of authors per paper. In the 1980s, the average number of authors per paper of hydrological journal articles was below 2 while this figure has soared to 4 to 5 in 2017, depending on the journal. While European Research Council (ERC) and other internationally funded research often necessarily involve multiple authorships, this does make an individual's contribution difficult to determine and advantages "networkers" as much as "true contributors". Although long author lists are evidently not negative per se, as they demonstrate the need for collaboration and integration of specialized knowledge, they may be problematic when used for research assessments. Koutsoyiannis et al. (2016) suggested addressing this issue by normalizing citation statistics by the number of authors. There have been similar discussions in other disciplines. In medicine, for example, a new approach to authorship transparency has been formalized through the CRediT (**C**ontributor **R**oles **T**axonomy) initiative (see http://docs.casrai.org/CRediT, last access: 2 November 2018, and McNutt et al., 2018). While the discipline reflects on ways of dealing with this challenge, we recommend that, in the interim, multi-authored research papers should include a statement of attribution of contributions, specifying who of the author list contributed in designing the research, conducting the research, writing the text, editing the text and funding the research. Furthermore, these trends are located within a changing landscape of academic publishing. Research funders and users of research outputs are increasingly demanding open access, and publishers are grappling with different models. This adds additional complexities to the issues of quantity, speed and multiple authorships.

3 Recognizing importance of novel insight

The main purpose of scientific publication consists of communicating new, important findings to peers in order to advance the science. The main role of editors, together with authors, reviewers and associate editors, is to maximize the potential towards fostering progress. During the publication process, the degree to which the manuscript contributes to advancing our science is in theory detected by the peer review system. However, as publications become more numerous, models more complex and data sets more extensive, it has sometimes become very difficult to assess the validity of a new theory or model prediction on the basis of the material contained in a manuscript. Most hydrology journals have therefore adopted a policy of open data and open models (e.g. Data Citation Synthesis Group, 2014), to allow peers – at least in principle – to repeat any published study. While the open data/model policies are recognized as being important, there are particular challenges in hydrology as, in some countries, the data (and models) used are often proprietary. Also, publication strategies often involve keeping part of the data for further analyses by the same group. Open data/model policies will certainly need particular attention in the near future and will likely require a change in the thinking of researchers and data collection agencies. Given the increasing burden that open data and open model policies impose on authors, institutions and journals should seek approaches that facilitate compliance.

A secondary purpose of scientific publication lies in recognizing the contributions of individuals and their research institutions. While, traditionally, this was done by attributing seminal achievements to the authors publishing them (e.g. Newton became famous through the power of ideas in his *Principia*), the process has today become more formalized due to the availability of publication databases and associated metrics. Typical assessment criteria are the number of publications, the citations they receive, and the quality of the journals in which they are published.

The quality of journals, as used in research assessments, is often quantified by journal impact factors (IFs). They are a measure of the number of citations to the papers of that journal over a particular period and have been used to separate reputable journals from low threshold web postings, new media and predatory journals (Beall, 2016). The presumption is that the quality of individual papers can somehow be inferred from the citation count of the journal as a whole. A comparison among six leading hydrology journals over the period 1996 to 2016, published as an editorial in *Water Resources Research* (Clark and Hanson, 2017), concludes that the journal impact factor in a given year does not have much predictive power for journal-level productivity. Impact factors, particularly in smaller journals, were found to vary substantially across years, which can be expected for statistical reasons (small samples). This is not to say that a journal's impact factor is not a useful metric; with many more journals appearing, an impact factor could be helpful, for example, in indicating journal development and maturity. The important point is that assessments of research quality and choices of journals for submitting work to should not be driven by impact factors. Furthermore, a comparison between disciplines suggests that the journal impact factors of hydrology journals are rather low (all journals reviewed have an impact factor of less than 5; Clark and Hanson, 2017) in relation to disciplines such as medicine, chemistry and physics, which highlights the problem of using impact factors to compare the quality of work across disciplines (Koutsoyiannis and Kundzewicz, 2007). In hydrology, papers tend to be cited over much longer time periods which, together with the smaller size of the discipline, means that the short 2-year time window for impact factor calculation is a limitation in our discipline. It is also influenced by the fact that impacts of some hydrological publications materialize through application to water-related management, which is not reflected in citations (Cudennec and Hubert, 2008).

It is arguable whether there is any set of metrics that would effectively measure a lasting contribution to academic thought and practice, quite apart from whether these could be gamed by an individual choosing to do so. A general concern therefore emerges from the current practice of assessing and ranking scientific productivity of institutions, journals and individuals by bibliometric indices which could indirectly incentivize academic misconduct (Edwards and Roy, 2017). We also note that the San Francisco Declaration on Research Assessment (DORA) (http://www.ascb.org/dora/, last access: 2 November 2018) urges a focus on the scientific contribution of published papers rather than where the papers were published in an attempt to reduce the misuse of impact factors for research assessment. Similarly, the commendable EU "Open Science" initiative and associated report on next generation responsible metrics (https://ec.europa.eu/research/openscience/pdf/report.pdf, last access: 2 November 2018) should inform our debate and practice. It would stand hydrology in good stead if we, like only a generation ago, assessed research impact (and the performance of individuals and institutions) by the changes in the thinking induced, rather than by citation numbers. This is the (unfortunately not objectively measurable) criterion that would maximize advances in science, suggesting that peer review assessments should be given higher priority in the future.

4 Role of journals in setting the science agenda

With climate change currently being high on the political agenda and coupled with prevailing publication pressures, it is not surprising that submissions on climate impact studies, often with little novelty or innovation, have become something of a cottage industry. Equally disappointing is the proliferation of model applications with marginal innovation and/or little generality. There is indeed an interesting question of whether societal needs, fundamental ideas or new technologies are the main drivers of scientific progress. Sivapalan and Blöschl (2017) suggested that all three have been and will be important ingredients in hydrology. They also noted that research progress has come about in discrete steps or "eras". For example, the two decades from 1970 to 1990 focused on hydrological processes involving substantial field work. Later the interest in field work ebbed away because of the high cost-to-benefit ratio (Blume et al., 2017) and changing societal priorities.

Indeed in the 21st century the human footprint is fast becoming a dominant feature in the hydrological cycle, and research across the disciplines is becoming mandatory. Publishing interdisciplinary research, however, still remains challenging. There is a tendency for researchers and their communities to be socialized within their own discipline niches, and communities may become self-reinforcing to the detriment of fresh outside perspectives. Most hydrology journals have already responded strategically to these interdisciplinary publication needs, for example, by selecting editors and reviewers from a diverse set of disciplines. The strategic response of *Water Resources Research (WRR)* is a potential approach to help mature interdisciplinary thinking. *WRR* encourages didactic reviews to provide the perspective of other disciplines (i.e. how they undertake research and engage discourse within their field) and also commentary papers that explore why a particular field is struggling and seeks to explore the field from multiple perspectives.

Whether the research is disciplinary or inter-/multi-disciplinary, journals play an important role in communicating and setting the trend for the vision of hydrological research, and for fostering innovation in a coherent way. We need to work collectively to ensure that science of the highest quality and that innovative content is published in our journals. To do this the hydrological community must redress research investment deficiencies and the publication biases that arise as a result of a lack of funding. Research agendas should not be so narrowly linked to today's problems, and we need to be bold in setting out the grand challenges of our discipline. For example, the International Association of Hydrological Sciences (IAHS), in collaboration with the Hydrology Divisions of EGU and AGU, has recently called for compiling a list of unsolved scientific problems in hydrology that would invigorate research in the 21st century (https://iahs.info/IAHS-UPH/, last access: 2 November 2018). The initiative has been motivated by David Hilbert's (1900) unsolved problems, which have greatly stimulated focused research in mathematics. The idea is that a similar list of problems could be identified by the hydrological community. For tangible progress to be made the problems should be framed so they

- ideally relate to observed phenomena and why they happen
- are universal (i.e. not only apply to one catchment or region)
- are specific (so there is hope they can be solved).

We commend this initiative and urge colleagues to contribute to shaping progress in hydrology.

5 Summary and concluding remarks

Hydrology, a traditionally integrative science with high societal relevance and geographic diversity, is perhaps an optimal place from which to launch the movement to reassert the academic spirit in a time where there is dramatic change in the way people learn, synthesize and interact with each other. Our community stands at the cusp of perhaps the greatest societal revolution in the democratization of access to resources and knowledge, as well as to the largest population the world has ever seen. These societal and technological changes have major effects on the publishing landscape. For hydrological journals there is a unique opportunity to learn through harnessing the energies of the moment to continue to improve our concept of the world and the role water plays in it.

- *Publication quantity, speed and multiple authorships.* Authors, reviewers and editors are encouraged to prioritize research quality over quantity. Discussions are currently under way to discourage unethical behaviour of authors, reviewers and editorial board members. Measures may involve a system for sharing information on ethical misconduct across hydrology, in addition to reinforcing the guidelines of COPE (Committee of Publication Ethics), to which our journals adhere. Authors are encouraged to make a personal judgement on whether fast-track findings may be more appropriately communicated through scientific journals or other communication forms. Similarly we must emphasize transparency in authorship contributions; multi-authored papers should include a statement of attribution of the individual contributions.

- *Recognizing importance of novel insight.* Most hydrology journals have adopted a policy of open data and open models, to allow peers to repeat any published study and fully appreciate the validity and novelty of the material. For these policies to be fully embraced, a change in culture will be required by both researchers and data collection agencies. The issue of research assessment on the basis of impact factors (the "tyranny of metrics") (Delzon et al., 2016) is symptomatic of a larger problem that we need to address and act on; the core values of transparency and peer review are the foundations of the scientific and social capital of our journals, and these principles, combined with embracing alternate and still-to-emerge media, will ensure that journals remain the trusted and authoritative communications outlets for compelling ideas for, and of, the future. We need to identify ways of ensuring that the value of hydrological journals continues to be recognized; we need to ensure that they are a primary and effective forum for furthering the science and practice of hydrology, and presenting solutions to challenging problems. We also need to ensure the focus of research assessments is on the scientific contributions of individual journal papers rather than on impact factors.

- *Role of journals in setting the science agenda.* Journals play an important contributory role – together with their parent organizations and associated conferences – in communicating and setting the trend for the vision of hydrological research, and for fostering innovation in a coherent way. Research agendas should be forward looking and not be narrowly linked to today's problems. There is a need for the discipline to work collectively to redress such funding and publication biases that consequently arise. We need to ensure that science of the highest quality and innovative content that facilitates and invigorates hydrological research is published in our journals.

As a hydrological community we are experiencing unprecedented challenges emerging from the rapidly changing science communication landscape. These challenges also represent an opportunity for a renaissance in the scope and societal impact of our discipline. As we engage with new modes of communication, we must remain vigilant to ensure top-quality science distinguishes our journals from the mass of unverified online information. The success of new measures for author transparency, for reducing scientometric bias, and for reinvigorating the hydrological science agenda depends on your participation and engagement. To realize this renaissance, we urge all to act in support of the issues raised in this editorial through activities within journal institutions, professional societies and the broader community of practice.

Acknowledgements. This editorial has benefited from the insightful critique of three reviewers – Dani Or, Murugesu Sivapalan and Ian Littlewood – and we would like to extend our collective thanks to these reviewers for their useful perspectives, comments and additions.

References

Beall, J.: Essential information about predatory publishers and journals, International Higher Education, 86 (Summer 2016), 2 pp., https://doi.org/10.6017/ihe.2016.86.9358, 2016.

Blöschl, G., Bárdossy, A., Koutsoyiannis, D., Kundzewicz, Z., Littlewood, I., Montanari, A., and Savenije, H.: Joint editorial – On the future of journal publications in hydrology, Hydrol. Res., 45, 515–518, https://doi.org/10.2166/nh.2014.006, 2014.

Blume, T., van Meerveld, I., and Weiler, M.: The role of experimental work in hydrological sciences–insights from a community survey, Hydrolog. Sci. J., 62, 334–337, https://doi.org/10.1080/02626667.2016.1230675, 2017.

Brossard, D. and Scheufele, D. A.: Science, new media, and the public, Science, 339, 40–41, https://doi.org/10.1126/science.1232329, 2013.

Clark, M. and Hanson, R.: Editorial – The citation impact of hydrology journals, Water Resour. Res., 53, 4533–4541, https://doi.org/10.1002/2017WR021125, 2017.

Cudennec, C. and Hubert, P.: The multi-objective role of HSJ in processing and disseminating hydrological knowledge, Hydrolog. Sci. J., 53, 485–487, https://doi.org/10.1623/hysj.53.2.485, 2008.

Data Citation Synthesis Group: Joint Declaration of Data Citation Principles, edited by: Martone, M., San Diego CA, FORCE11, 2014.

Delzon, S., Cochard, H., and Pfautsch, S.: Indexing the indices: scientific publishing needs to undergo a revolution, Journal of Plant Hydraulics, 3, e009, https://doi.org/10.20870/jph.2016.e009, 2016.

Edwards, M. A. and Roy, S.: Academic research in the 21st century: Maintaining scientific integrity in a climate of perverse incentives and hypercompetition, Environ. Eng. Sci., 34, 51–61, https://doi.org/10.1089/ees.2016.0223, 2017.

Hilbert, D.: Mathematische Probleme, Nachrichten von der Königlichen Gesellschaft der Wissenschaften zu Göttingen – Mathematisch-Physikalische Klasse, 253–297, 1900.

Koutsoyiannis, D. and Kundzewicz, Z.: Editorial – Quantifying the impact of hydrological studies, Hydrolog. Sci. J., 52, 3–17, https://doi.org/10.1623/hysj.52.1.3, 2007.

Koutsoyiannis, D., Blöschl, G., Bárdossy, A., Cudennec, C., Hughes, D., Montanari, A., Neuweiler, I., and Savenije, H.: Joint editorial – Fostering innovation and improving impact assessment for journal publications in hydrology, Hydrolog. Sci. J., 61, 1170–1173, https://doi.org/10.1002/2016WR018895, 2016.

Martin, B.: Whither research integrity? Plagiarism, Self-plagiarism and coercive citation in an age of research assessment, Res. Policy, 42, 1005–1014, https://doi.org/10.1016/j.respol.2013.03.011, 2013.

McNutt, M., Bradford, M., Drazen, J., Hanson, B., Howard, B., Jamieson, K., Kiermer, V., Marcus, E., Pope, B., Scheckman, R., Swaminathan, S., Stang, P., and Verma, I.: Transparency in authors' contributions and responsibilities to promote integrity in scientific publication, P. Natl. Acad. Sci. USA, 115, 2557–2560, https://doi.org/10.1073/pnas.1715374115, 2018.

Sivapalan, M. and Blöschl, G.: The growth of hydrological understanding: Technologies, ideas, and societal needs shape the field, Water Resour. Res., 53, 8137–8146, https://doi.org/10.1002/2017WR021396, 2017.

Proc. IAHS, 380, 9–15, 2018
https://doi.org/10.5194/piahs-380-9-2018

Open Access
PIAHS

Monitoring environmental supporting conditions of a raised bog using remote sensing techniques

Saheba Bhatnagar, Bidisha Ghosh, Shane Regan, Owen Naughton, Paul Johnston, and Laurence Gill

Trinity College Dublin, Department of Civil, Structural and Environmental Eng, Dublin, Ireland

Correspondence: Saheba Bhatnagar (sbhatnag@tcd.ie)

Received: 14 April 2018 – Revised: 24 September 2018 – Accepted: 25 September 2018 – Published: 18 December 2018

Abstract. Conventional methods of monitoring wetlands and detecting changes over time can be time-consuming and costly. Inaccessibility and remoteness of many wetlands is also a limiting factor. Hence, there is a growing recognition of remote sensing techniques as a viable and cost-effective alternative to field-based ecosystem monitoring. Wetlands encompass a diverse array of habitats, for example, fens, bogs, marshes, and swamps. In this study, we concentrate on a natural wetland – Clara Bog, Co. Offaly, a raised bog situated in the Irish midlands. The aim of the study is to identify and monitor the environmental conditions of the bog using remote sensing techniques. Environmental conditions in this study refer to the vegetation composition of the bog and whether it is in an intact (peat-forming) or degraded state. It can be described using vegetation, the presence of water (soil moisture) and topography. Vegetation indices (VIs) derived from satellite data have been widely used to assess variations in properties of vegetation. This study uses mid-resolution data from Sentinel-2 MSI, Landsat 8 OLI for VI analysis. An initial study to delineate the boundary of the bog using the combination of edge detection and segmentation techniques namely, entropy filtering, canny edge detection, and graph-cut segmentation is performed. Once the bog boundary is defined, spectra of the delineated area are studied. VIs like NDVI, ARVI, SAVI, NDWI, derived using Sentinel-2 MSI and Landsat 8 OLI are analysed. A digital elevation model (DEM) was also used for better classification. All of these characteristics (features) serve as a basis for classifying the bog into broad vegetation communities (termed "ecotopes") that indicate the quality of raised bog habitat. This analysis is validated using field derived ecotopes. The results show that, by using spectral information and vegetation index clustering, an additional linkage can be established between spectral RS signatures and wetland ecotopes. Hence, the benefit of the study is in understanding ecosystem (bog) environmental conditions and in defining appropriate metrics by which changes in the conditions can be monitored.

1 Introduction

A bog is a type of wetland which primarily depends on rainfall for water and nutrients. Bogs can be categorised as blanket bog and raised bog. Raised bogs are discrete, raised, dome-shaped masses of peat occupying former lakes or shallow depressions in the landscape (Fossitt, 2000). They occur throughout the midlands of Ireland (Felicity Hayes-McCoy, 2017) and in this study, we focus on one of the largest raised bogs in Ireland, Clara Bog, Co. Offaly. Monitoring wetland structure and function typically requires recurrent site visits, which can be prohibitively labour intensive, costly and time-consuming. Monitoring is often unfeasible due to the poor accessibility, and is thus, only practical on relatively small areas (Adam et al., 2010). To acquire frequent measurements and timely information remote sensing (RS) is a cost-effective tool. Remote sensing provides invaluable information to characterize and measure the conditions of wetlands and their functioning.

The current state of art primarily focuses on mapping different types of the wetlands (Mahdavi et al., 2017). The mapping is done using different wavelengths and spectral response of the objects. Satellite imagery-derived vegetation indices can be effectively used for assessing the vegetation status of an ecosystem. Vegetation communities present within an ecosystem are defined as ecotopes. The Normalized Difference Vegetation Index (NDVI), Soil Adjusted

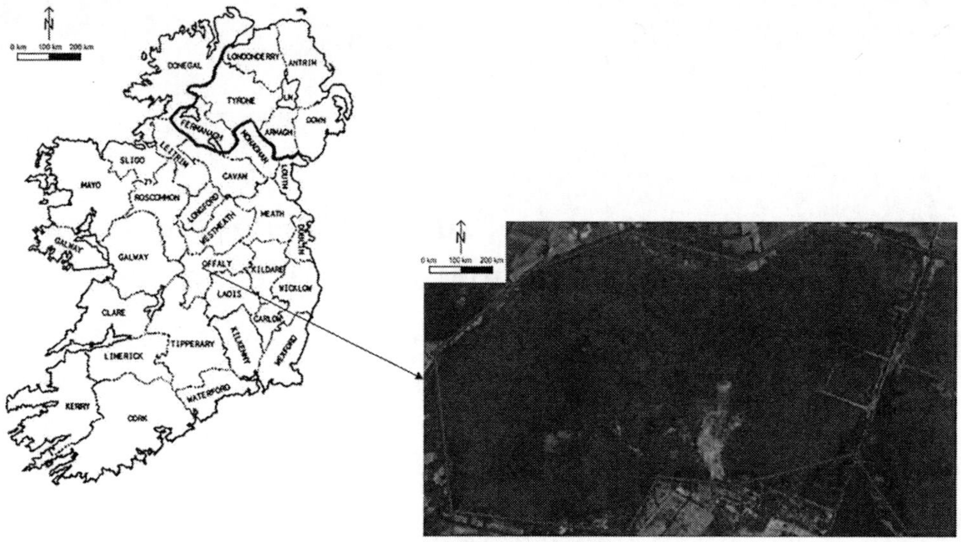

Figure 1. Clara Bog, Co. Offaly.

Vegetation Index (SAVI), Atmospherically Resistant Vegetation Index (ARVI) are the most effective vegetation indices stated in the literature (Wiegand et al., 1991). Soil moisture is not directly derivable from the optical bands. The Normalized Difference Water Index (NDWI) using near-infrared (NIR) and short-wave infrared (SWIR) bands give an indication of wetness of the surface inferred as soil moisture. These indices can be used to provide a clearer picture of vegetation and water extent in an area.

Topography plays a vital role in analysing an ecosystem. It gives an accurate idea of elevation difference present between various plant communities. Light detection and ranging (LiDAR) provides with point cloud information which can be used to deduce the topography of an area. LiDAR systems can be terrestrial, airborne, or space-borne. Normally, terrestrial systems are used for 3-D reconstruction, whereas air and space systems are utilized for remote sensing and wide territory mapping. In this study, an airborne LiDAR derived DEM is used for analysing the topographic extent of the bog.

For analysis of the RS data, various machine learning tools have proven to be useful (Lu and Weng, 2007). There are many state of the art segmentation and classification algorithms available. It is necessary to make full use of the advantages of different algorithms on the basis of multi-feature fusion, so as to achieve better segmentation effect (Yuheng and Hao, 2017). Hence, in this study a combination of segmentation algorithms is deployed. Classification accuracy is tested using various classifiers namely, SVM, Bagged Tree and Subspace KNN. SVM can be tuned using the value of optimization parameter, kernel used and hence, overfitting can be avoided. Bagged Tree, Subspace KNN are ensemble classifiers. The idea behind ensemble classifiers is to learn from a set of classifiers rather than a single classifier. The final result is either the average of the result from all the classifiers or is obtained using majority voting. An ensemble learner is more robust and less manpower is required for tuning the parameters. Here, a comparative study on the performance of the classifiers is carried out using freely available Landsat 8 OLI and Sentinel 2 MSI data for monitoring ecological condition and mapping ecotopes present inside the bog.

2 Materials and methodology

2.1 Study region and datasets

The site selected for this study is one of the largest bogs in Ireland, Clara Bog (Fig. 1), covering approximately 840 ha of which 443.36 ha is uncut high bog with the remaining 393.18 ha mostly cutover bog (About Clara Bog, 2018).

In the Clara bog, 9 broad categories of ecotopes have been defined namely, Sub-marginal, Sub-central, Marginal, Central, Inactive flush, Active flush, Open water, Face bank and Bog woodland.

The health of the bog is indicated by its ability to form peat. The formation of peat is depicted by central, subcentral and active flush. Ecotopes like marginal, submarginal are indicative of peatland degradation (Fernandez Valverde, 2012).

For the best description of the bogs following open source data is used:

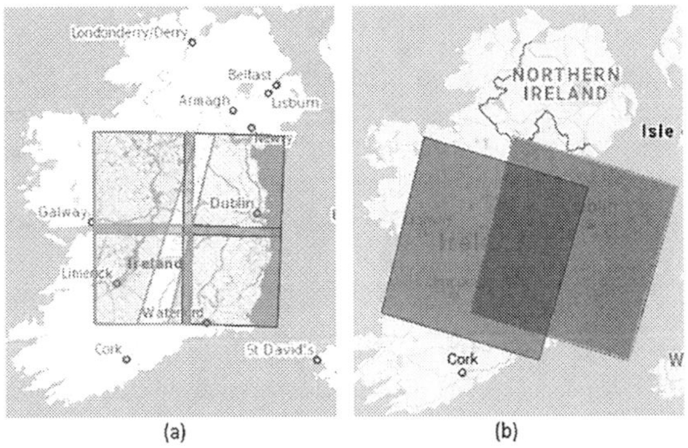

Figure 2. Footprints over Clara, Co. Offaly, Ireland **(a)** Sentinel-2, L2A, Tile_Id – 29UNV (https://scihub.copernicus.eu, last access: 5 March 2018) **(b)** Landsat-8, OLI+TIRS, Path 207, Row 023 (https://earthexplorer.usgs.gov/, last access: 19 February 2018)

1. Sentinel-2 Multispectral Instrument Level 2A (S2-MSIL2A).

   S2MSIL2A has bottom-of-atmosphere (BOA) reflectance in cartographic geometry. The granules also called tiles, are $100 \times 100\,\text{km}^2$ ortho-images in UTM/WGS84 projection. The L2A-BOA product is atmospherically corrected and ready to use (Gatti and Bertolini, 2013) and is accessed from https://scihub.copernicus.eu/ (last access: 5 March 2018). The area at test lies under tile id – T29UNV. Sentinel-2 has total of 12 bands out of which 9 bands are used for analysis in this study (Band 2–8A, Band 11).

2. Landsat 8 Combined (LC08).

   Landsat 8 carries two push-broom instruments: The Operational Land Imager (OLI) and the Thermal Infrared Sensor (TIRS) in UTM/WGS84 projection. Atmospherically corrected, ready to use data is accessed from http://earthexplorer.usgs.gov/ (last access: 19 February 2018). The area at test lies under path 207, row 23. In this study, 12 bands are used for analysis (Band 1–11 + Pixel quality assessment, QA) (Zanter, 2016).

The latitde-longitude extent of the Clara bog is 53°19′47″ N, 7°39′34″ W; 53°18′55″ N, 7°37′24″ W. Images used are acquired by S2 and L8 on the same date (20 June 2017). Image from Sentinel-2 is resampled to 10 m (appropriate bands) and image from Landsat 8 is resampled to 30 m. The footprints for both the satellites can be seen in Fig. 2.

2.2 Methodology

The methodology used is described in the following flowchart (Fig. 3). First, it is necessary to delineate the ecosystem from the surrounding area in order to minimise

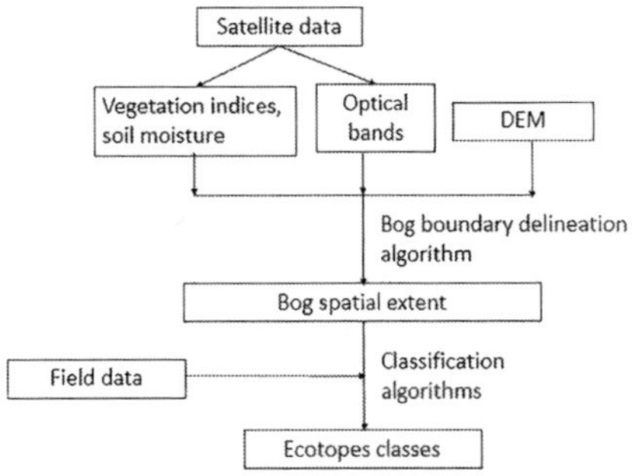

Figure 3. Methodology Flowchart.

the effect of outliers; this is achieved using segmentation algorithms. The delineated ecosystem is further divided into vegetation communities or ecotopes using a set of ensemble classifiers, namely Bagged Tree and Subspace KNN along with SVM.

2.3 Vegetation Indices and Soil Moisture

The vegetation indices used in this study are (Vegetation Indices, 2018):

1. Normalized Difference Vegetation Index

   $$\text{NDVI} = (\text{NIR} - \text{Red}) / (\text{NIR} + \text{Red}) \quad (1)$$

   NDVI indicates the amount of vegetation, distinguishes vegetation from the soil, minimizes topographic effects, etc.

2. Soil Adjusted Vegetation Index

   $$\text{SAVI} = [(\text{NIR} - \text{Red}) / (\text{NIR} + \text{Red} + L)] \cdot (1 + L) \quad (2)$$

   where L is a soil correction factor.

3. Atmospherically Resisted Vegetation Index

   $$\text{ARVI} = (\text{NIR} - \text{RB}) / (\text{NIR} + \text{RB}) \quad (3)$$

   where RB is a combination of the reflectance in the Blue (B) and Red (R) channels $\text{RB} = \text{R} - \gamma(\text{B} - \text{R})$ and γ depends on the aerosol type.

4. Soil Moisture: Normalized Difference Water Index

   $$\text{NDWI} = (\text{NIR} - \text{SWIR}) / (\text{NIR} + \text{SWIR}) \quad (4)$$

   Values range from -1, very low moisture level, to 1 very high moisture level.

Hence, there are a total of 5 extra layers i.e., NDVI, SAVI, ARVI, NDWI, and DEM along with satellite bands which are fed into the algorithm as input characteristics. Therefore, for Sentinel-2 data set there is a total of 14 layers and for Landsat 8 there is a total of 17 layers.

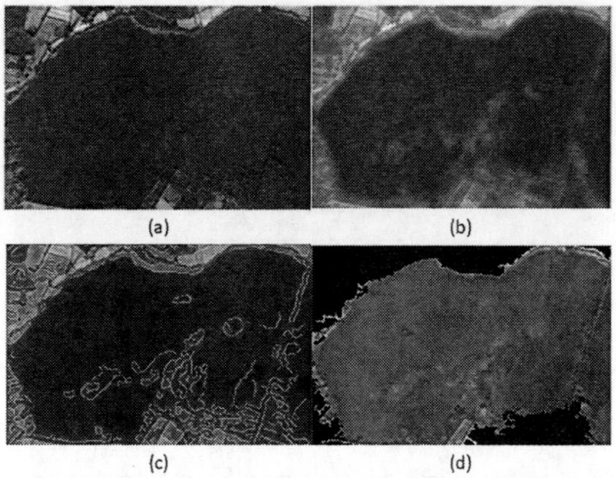

Figure 4. **(a)** Original Clara Image **(b)** Entropy Filter Image **(c)** Canny Edge Image **(d)** Boundary Delineated Image.

2.4 Bog Boundary Delineation

Delineation of the wetland extent was carried out using three algorithms in conjunction:

1. Entropy Filtering: measures the relative change in entropy for detection of edges. All the areas with potential objects are thus highlighted.

2. Canny Edge Detection: Initially, the intensity gradient is measured according to which background pixels are removed, i.e. only thin lines depicting edges remain. The algorithm uses 2 thresholds (upper and lower) to accept the pixels as the edge.

3. Graph Cut Segmentation: Divides every pixel into foreground (source) and background (sink) based on probability and neighbourhood information. Two major steps then follow, first is to construct the graph and the second is to produce min cut (i.e. max flow).

Since the ecosystem and surrounding areas contain distinct vegetation, the NDVI image is used as the base image (Bi) for defining the bog boundary for Sentinel data, and band 8 (panchromatic) for Landsat 8 dataset.

Algorithm 1 Bog Boundary Delineation

$EBi \leftarrow$ Entropy_filter(Bi)
$CBi \leftarrow$ CannyEdge_Detect(Bi, EBi)
$GCi \leftarrow$ Graph_Cut(Bi, CBi)

Graph Cut image (GCi) is the delineated image. From GCi, spurious regions are removed on the basis of thresholding (Fig. 4).

2.5 Ecotope Identification and Classification

In this study, we have explored the applicability of pixel based, supervised classification on a raised bog. The classifier is first trained on a subset of available data, and then tested on a new location (test) to predict the classes (ecotopes) present. Here, we are doing a direct transfer of pixel-based knowledge from the training area to the testing area. The following classifiers are used for this purpose:

1. SVM – These are supervised learning models with associated learning algorithms that analyse data used for classification and regression analysis. Given a set of training data, an SVM training algorithm builds a model that assigns new data to one of the two categories, making it a non-probabilistic binary linear classifier (Cortes and Vapnik, 1995).

2. Bagged Tree – Ensemble, supervised classifier. It approaches to combine several machine learning techniques into one predictive model in order to decrease the variance hence, tuning the prediction into an expected outcome (Ensembles, 2018).

3. Subspace KNN – Similar to Bagging, subspace KNN is an ensemble method to reduce the correlation between estimators (Ho, 1998).

2.6 Validation

In this study, five major classes are considered namely Sub-marginal, Sub-central, Marginal, Central and Active flush as these are the key ecological classes indicating bog condition.

The result achieved is verified using field derived ecotope map (Fig. 5). Initially, a classification model is created using training data. For each of the classifier and data set, Model Accuracy (MA), Transfer Accuracy (TA) and Kappa Coefficient (kappa) is measured. The model accuracy (MA) is measured using 5-fold validation of training data. The transfer accuracy (TA) is the test accuracy when the model is applied to the testing data.

Ground truth is divided into:

- Case 1: Training (50 %) and Testing (50 %)

- Case 2: Training (70 %) and Testing (30 %)

3 Results and discussion

The results obtained using the aforementioned algorithms are validated using field derived ecotopes. The spatial location of the ecotopes is not weather dependent, compared to satellite imagery and corresponding vegetation indices which change with respect to weather and other environmental conditions. This study primarily highlights the condition of the raised bog during the summer season. TA signifies the scope of

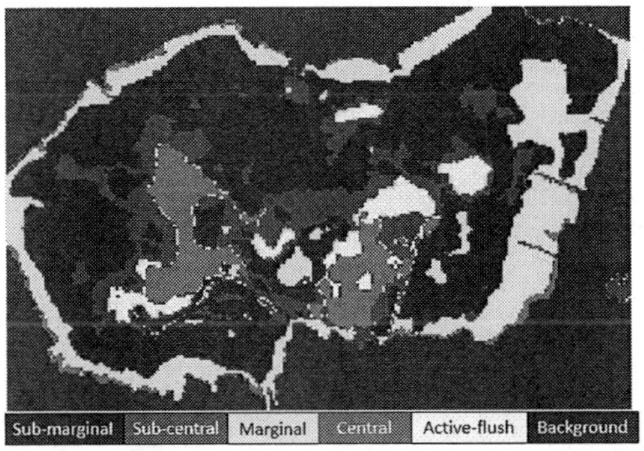

Figure 5. Ground truth with 5 ecotopes.

Table 1. Accuracies for all the cases.

	MA %	TA %	kappa
Sentinel-2 Case 1:			
SVM	73.96	32.40	0.1302
Bagged Tree	83.63	53.46	0.3424
Subspace KNN	81.65	52.95	0.3370
Sentinel-2 Case 2:			
SVM	71.71	36.36	0.2050
Bagged Tree	83.38	65.23	0.5013
Subspace KNN	62.02	36.50	0.1922
Landsat-8 Case 1:			
SVM	77.82	26.76	0.1305
Bagged Tree	87.27	53.87	0.3883
Subspace KNN	87.22	51.97	0.3635
Landsat-8 Case 2:			
SVM	72.92	29.06	0.1711
Bagged Tree	85.55	52.69	0.3539
Subspace KNN	85.00	52.71	0.3549

transferring the knowledge gained from the first half (Train) to identify ecotopes in the second half (Test).

Table 1 states the accuracy achieved using the aforementioned methodology:

3.1 Discussion

In this study, we have studied the ecological conditions of a raised bog using data from Sentinel-2, Landsat-8, their vegetation derivatives, and DEM. The key points in the result are discussed below:

1. The use of a single algorithm for boundary delineation leads to the formation of smaller, non-connected objects and hence, the bog is not delineated properly. Using the entropy filter, canny edge detection, and Graph Cut in conjunction proved to be an effective way of delineating a complex structure from a middle-resolution image.

Table 2. Number of pixels per class – S2.

Name of Ecotope	Number of pixels
Submarginal	13 771
Subcentral	4633
Marginal	4913
Central	1192
Active Flush	2443
Background	12 248

2. SVM achieved the highest accuracy in the delineation of the submarginal ecotope, but was not viable for marginal, active flush or background. The classifier has confused between marginal, active flush and background, giving 0 % class accuracy in both datasets (Fig. 6). SVM has a major drawback of tuning the parameters. In this study the parameters were kept constant for both datasets, which increased the chances of overfitting, hence, the low accuracy.

3. Ensemble classifiers (BT, SKNN) show similar results due to the fact that the ensemble methods are generally consistent (in terms of their effect on accuracy) (Figs. 7, 8) (Maclin and Opitz, 2011).

4. The test (OA) accuracy (transfer) is also highly dependent on the number of training pixels (Table 2).

   Total pixels in the image (Sentinel 2 MSI; $160 \cdot 245) = 39\,200$. The Submarginal ecotope is most correctly classified by all the classifiers followed by Marginal. Since other ecotopes are present in much lower quantities (compared to submarginal and background), they are not identified correctly.

5. Similarity between the signatures of the classes:

   Jeffries-Matusita (JM) distance is a widely used method for feature selection in multiclass problems (Swain and Davis, 1978). The values of JM distance (Table 3) between the ROI pixels depicts low spectral separability between the ecotopes. These values were measured using 100 points from each ecotope-pair. A lower value means low separability and higher value shows higher spectral separability between ecotope pair. A higher value of JM distance is desirable for better identification of classes (Whelley et al., 2014).

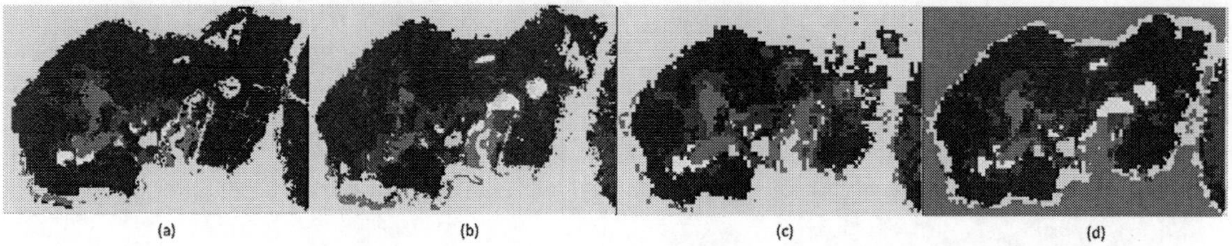

Figure 6. SVM Classified Image: **(a)** Sentinel-2 Case 1 **(b)** Sentinel-2 Case 2 **(c)** Landsat-8 Case 1 **(d)** Landsat-8 Case 2.

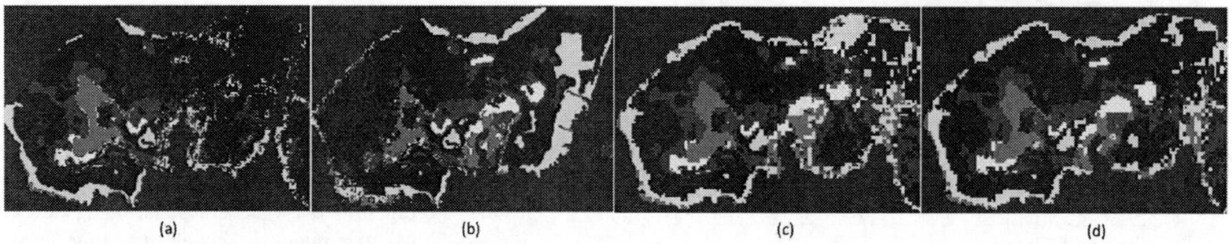

Figure 7. Bagged Tree Classified Image: **(a)** Sentinel-2 Case 1 **(b)** Sentinel-2 Case 2 **(c)** Landsat-8 Case 1 **(d)** Landsat-8 Case 2.

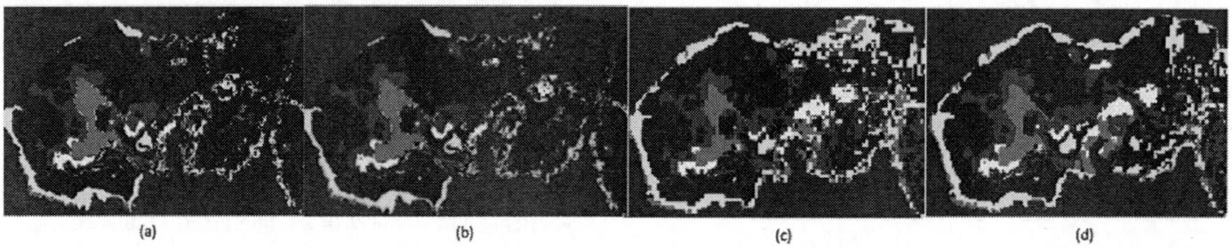

Figure 8. Subspace KNN Classified Image: **(a)** Sentinel-2 Case 1 **(b)** Sentinel-2 Case 2 **(c)** Landsat-8 Case 1 **(d)** Landsat-8 Case 2.

Table 3. JM distance between ecotopes

Ecotope Pair	JM distance
Submarginal and Subcentral	0.22
Submarginal and Marginal	0.85
Subcentral and Central	0.67
Central and Activeflush	1.48

4 Conclusions

In this study, we studied the application of mid-resolution satellite data for classification of a raised bog. The study was carried out using data from two satellites, Sentinel-2 and Landsat-8. The data used was resampled to 10 m for appropriate Sentinel-2 bands and 30 m for Landsat-8. The final classification accuracy was similar for both the satellite, unaffected by the resolution of the images. The study initially describes a competent way of boundary delineation using a series of edge detection techniques. Vegetation indices along with soil moisture and DEM information are used as features to train the classification algorithms. Bagged tree (BT) classifier proves to be the best classifier for classification of the raised bog providing better accuracy than SVM or SKNN. This is due to the nature of the ensemble classifier to reduce variance and avoid overfitting. Transfer of knowledge directly from train location to test location is not achieved effectively due to limitations in data-resolution and the amount of input training pixels. The study suggests that transfer of knowledge is effective between similar ecosystems when there is a distinct difference in the distribution of various ecotopes and pixels can be unmixed.

Data availability. The data for this study is taken from Landsat-8 and Sentinel-2, and is openly available to use. Landsat-8 (USGS EarthExplorer, retrieved 19 February 2018 from https://earthexplorer.usgs.gov/), Sentinel-2 (Copernicus Open Access Hub, retrieved 5 March 2018 from https://scihub.copernicus.eu/). Ecotope data was provided by the National Parks and Wildlife Service as vector ESRI shapefiles and is available upon request.

Author contributions. SB and BG conceived of the presented idea. SB developed the theory and performed the computations, and BG and ON verified the analysis. The verification data was provided by SR, SB, BG, ON, SR, LG and PJ contributed to the design of the research and to the writing of the manuscript.

Competing interests. The authors declare that they have no conflict of interest.

Special issue statement. This article is part of the special issue "Earth Observation for Integrated Water and Basin Management: New possibilities and challenges for adaptation to a changing environment". It is a result of The Remote Sensing & Hydrology Symposium, Cordoba, Spain, 8–10 May 2018.

Acknowledgements. This study is funded by the Environmental Protection Agency of Ireland. The authors would like to thank the National Parks and Wildlife Service for providing all ecological data and advice.

Edited by: Michael Cosh
Reviewed by: two anonymous referees

References

About Clara Bog: http://raisedbogs.ie/about-clara-bog/, last access: 28 March 2018.

Adam, E., Mutanga, O., and Rugege, D.: Multispectral and hyperspectral remote sensing for identification and mapping of wetland vegetation: a review, Wetl. Ecol. Manag., 18, 281–296, https://doi.org/10.1007/s11273-009-9169-z, 2010.

Cortes, C. and Vapnik, V.: Support-vector networks, Machine Learning, 20, 273–297, https://doi.org/10.1007/BF00994018, 1995.

Ensembles: available at:https://martin-thoma.com/ensembles/, last access: 8 March 2018.

Felicity Hayes-McCoy, W. J.: Dingle and its Hinterland: People, Places and Heritage, The Collins Press, 224 pp., 2017.

Fernandez Valverde, F., Crowley, W., and Wilson, S.: Raised Bog Monitoring Project 2011, in: Volume 1: Main Report, Irish Wildlife Manuals No. 62, National Parks and Wildlife Service, Department of Arts, Heritage and the Gaeltacht, Dublin, Ireland, 2012.

Fossitt, J. A.: A Guide to Habitats in Ireland, The Heritage Council, Co Wicklow, Ireland, 2000.

Gatti, A. and Bertolini, A.: Sentinel-2 products specification document, available at: https://earth.esa.int/documents/247904/685211/Sentinel-2+Products+Specification+Document (last access: 5 March 2018), 2013.

Ho, T. K.: Nearest neighbors in random subspaces, in: Advances in Pattern Recognition, edited by: Amin, A., Dori, D., Pudil, P., and Freeman, H., 640–648, Springer Berlin Heidelberg, Berlin, Heidelberg, 1998.

Lu, D. and Weng, Q.: A survey of image classification methods and techniques for improving classification performance, Int. J. Remote Sens., 28, 823–870, https://doi.org/10.1080/01431160600746456, 2007.

Maclin, R. and Opitz, D. W.: Popular Ensemble Methods: An Empirical Study, CoRR, abs/1106.0257, http://arxiv.org/abs/1106.0257 (last access: 16 March 2018), 2011.

Mahdavi, S., Salehi, B., Granger, J., Amani, M., Brisco, B., and Huang, W.: Remote sensing for wetland classification: a comprehensive review, GISci. Remote Sens., 55, 623–658, https://doi.org/10.1080/15481603.2017.1419602, 2017.

Swain, P. H. and Davis, S. M.: Remote sensing: the quantitative approach, IEEE T. Pattern Anal. Mach. Intel., 1, 713–714, 19784.

Vegetation Indices: available at: http://web.pdx.edu/~nauna/resources/8-2012_lecture1-vegetationindicies.pdf, last access: 6 March 2018.

Wiegand, C. L., Richardson, A. J., Escobar, D. E., and Gerbermann, A. H.: Vegetation indices in crop assessments, Remote Sens. Environ., 35, 105–119, 1991.

Whelley, P. L., Glaze, L. S., Calder, E. S., and Harding, D. J.: LiDAR-Derived Surface Roughness Texture Mapping: Application to Mount St. Helens Pumice Plain Deposit Analysis, IEEE T. Geosci. Remote, 52, 426–438, https://doi.org/10.1109/TGRS.2013.2241443, 2014.

Yuheng, S. and Hao, Y.: Image Segmentation Algorithms Overview, CoRR, abs/1707.02051, http://arxiv.org/abs/1707.02051 (last access: 19 March 2018), 2017.

Zanter, K.: Landsat 8 (L8) data users handbook, Landsat Science Official Website, available at: https://landsat.usgs.gov/landsat-8-l8-data-users-handbook (last access: 19 February 2018). 2016.

Proc. IAHS, 380, 17–22, 2018
https://doi.org/10.5194/piahs-380-17-2018

Open Access
PIAHS

Evapotranspiration and evaporation/transpiration partitioning with dual source energy balance models in agricultural lands

Gilles Boulet[1], Emilie Delogu[1], Sameh Saadi[1,2], Wafa Chebbi[1,2], Albert Olioso[3], Bernard Mougenot[1], Pascal Fanise[1], Zohra Lili-Chabaane[2], and Jean-Pierre Lagouarde[4]

[1]CESBIO, Université de Toulouse, CNES/CNRS/IRD/UPS, Toulouse, France
[2]Université de Carthage/Institut National Agronomique de Tunisie, Tunis, Tunisie
[3]EMMAH, INRA, Université d'Avignon et des Pays de Vaucluse, Avignon, France
[4]ISPA, INRA, Bordeaux Sciences Agro, Villenave d'Ornon, France

Correspondence: Gilles Boulet (Gilles.Boulet@ird.fr)

Received: 23 April 2018 – Revised: 10 October 2018 – Accepted: 15 October 2018 – Published: 18 December 2018

Abstract. EvapoTranspiration (ET) is an important component of the water cycle, especially in semi-arid lands. Its quantification is crucial for a sustainable management of scarce water resources. A way to quantify ET is to exploit the available surface temperature data from remote sensing as a signature of the surface energy balance, including the latent heat flux. Remotely sensed energy balance models enable to estimate stress levels and, in turn, the water status of most continental surfaces. The evaporation and transpiration components of ET are also just as important in agricultural water management and ecosystem health monitoring. Single temperatures can be used with dual source energy balance models but rely on specific assumptions on raw levels of plant water stress to get both components out of a single source of information. Additional information from remote sensing data are thus required, either something specifically related to evaporation (such as surface water content) or transpiration (such as PRI or fluorescence). This works evaluates the SPARSE dual source energy balance model ability to compute not only total ET, but also water stress and transpiration/evaporation components. First, the theoretical limits of the ET component retrieval are assessed through a simulation experiment using both retrieval and prescribed modes of SPARSE with the sole surface temperature. A similar work is performed with an additional constraint, the topsoil surface soil moisture level, showing the significant improvement on the retrieval. Then, a flux dataset acquired over rainfed wheat is used to check the robustness of both stress levels and ET retrievals. In particular, retrieval of the evaporation and transpiration components is assessed in both conditions (forcing by the sole temperature or the combination of temperature and soil moisture). In our example, there is no significant difference in the performance of the total ET retrieval, since the evaporation rate retrieved from the sole surface temperature is already fairly close to the one we can reconstruct from observed surface soil moisture time series, but current work is underway to test it over other plots.

1 Introduction

There is an increasing need for spatially distributed estimates of agricultural water needs and therefore evapotranspiration (ET). Estimating evapotranspiration, and, in turn, water stress, is important for irrigation monitoring and drought assessment. To do so, Remote Sensing provides an important array of data and solutions. Three spectral domains are concerned: solar (Visible/Near InfraRed spectrum, e.g. NDVI), thermal (Thermal InfraRed, e.g. surface temperature) and microwave (Radar data mostly). NDVI quantifies the amount of green vegetation, the largest water user in most areas since plants assess a larger fraction of the soil water through roots than what contributes to evaporation. Surface temperature is related to water stress through the energy budget, and gives a clue about the difference between actual and potential ET

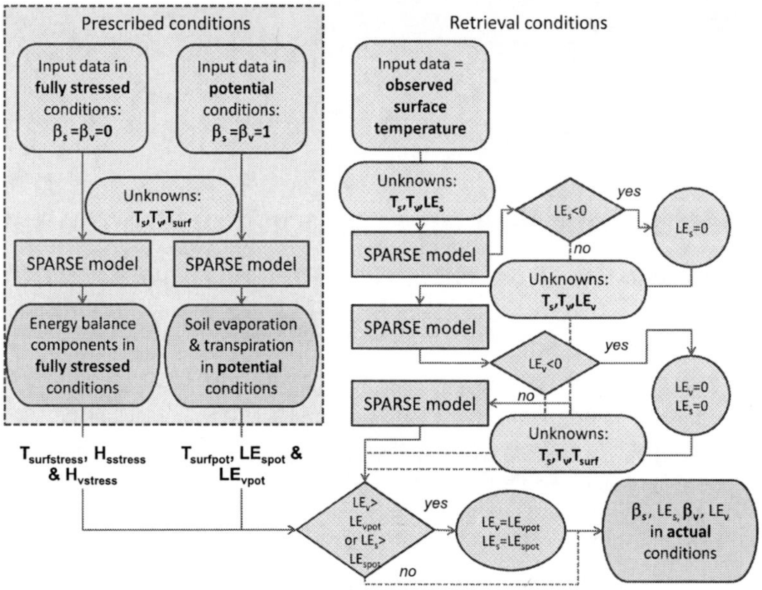

Figure 1. Flowchart of the SPARSE model (T is the element skin temperature, T_{surf} is the surface radiative temperature, LE is the latent heat flux and H the sensible heat flux, subscript "s" for soil and "v" for vegetation are used to characterize the component fluxes, subscripts "stress" for stressed and "pot" for potential are used to describe the water status; β is the efficiency, i.e. the ratio between actual and potential latent heat fluxes; from Saadi et al. (2018).

rates. Finally, radar is related to surface soil moisture and thus evaporation. While NDVI and radar, on the one hand, and NDVI and surface temperature, on the other, are frequently used together to estimate ET, the three sources of information have rarely been combined together.

ET is interesting for water management, drought assessment and irrigation control (esp. for drip or complementary irrigation), but one must also estimate separately evaporation and transpiration (the later represents the plant water uptake and the ecoagrosystem health). An estimate of the separate contribution of E and T to ET can be deduced from dual-source energy balance models such as TSEB (Kustas et al., 1999) or SPARSE (Boulet et al., 2015), but retrieving two unknowns (E and T) out of a single source of information (surface temperature T_{surf}) means that an additional assumption is laid down. In TSEB or SPARSE, the initial guess on the plant water status is that, in most cases, there is no stress, and T_{surf} is used to estimate E while T is computed by solving the plant energy budget in potential (i.e. unstressed) conditions. If the vegetation is suffering from water stress, its temperature will be higher than what is deduced from the energy budget in potential conditions. Consequently, the soil temperature that corresponds to the observed surface temperature and the underestimated vegetation temperature will be overestimated, and at some point this leads to a negative E retrieval. In that case TSEB and SPARSE assume that, if the vegetation is suffering from stress, the soil surface is already long dry, and E is close to zero. T_{surf} is thus used to retrieve T. But how robust is this? Can we improve the robustness by forcing E and T by two RS data, T_{surf} and a relative soil moisture level deduced from radar data (Schmugge et al., 1980)? This is the purpose of the present paper. It is organized in 3 main sections: the first summarize the retrieval and prescribed algorithms of SPARSE. The second is a numerical experiment assessing the limits of E and T retrievals in many configurations when using the sole surface temperature. The third section presents how forcing by both T_{surf} and a relative soil moisture level affects the retrieval processes for a real case study.

2 The retrieval/prescribed algorithms of SPARSE

SPARSE solves the dual-source energy budget of the soil and the vegetation. The model can be run in two modes: a retrieval mode to simulate evaporation and transpiration from TIR data, and a prescribed mode which simulates evaporation and transpiration rates for known stress levels (from fully stressed, i.e. $E = T = 0$ to fully potential). This enables to simulate not only actual fluxes but also surface and plant water stress. The prescribed (or direct) mode simulates fluxes and component (soil and vegetation) temperatures from known water stress conditions corresponding from any level between unstressed (potential rate) to fully stressed

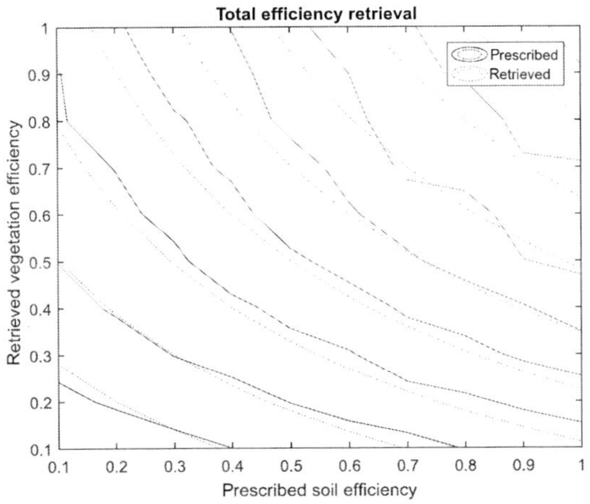

Figure 2. Isolines of total efficiency β simulated for each (β_s, β_V) combination using the model in prescribed (continuous lines) or retrieval (dotted lines) modes; the surface temperature generated by the model in prescribed mode for a particular (β_s, β_V) is used as input for the retrieval mode.

(minimum ET). The retrieval (or inverse) mode infer E and T from surface temperature observations using a decision tree (Fig. 1). In what follows, we use the complementary part to one of the stress, also named "efficiency", to characterize the relative stress levels: soil evaporation efficiency β_s is the ratio between the actual and potential soil evaporation rates, transpiration efficiency β_V is the ratio between the actual and potential transpiration rates and the total efficiency β is the ratio between the actual and potential total evapotranspiration rates. Here, we focus on instantaneous latent heat fluxes in W m^{-2} at the satellite overpass time instead of daily E and T values.

3 Synthetic experiment

3.1 "Classical" configuration: using the sole surface temperature data as input for flux retrieval

In order to infer the limits of the E/T retrieval, a synthetic study was carried out. In Fig. 1, the model is run for the two limiting cases (fully stressed, i.e. $\beta_s = \beta_V = 0$, or fully potential, $\beta_s = \beta_V = 1$). The model in prescribed mode can also be run for any combination of relative stress levels for the soil ($0 < \beta_s < 1$) and the vegetation ($0 < \beta_V < 1$) and produce the corresponding component fluxes and equilibrium surface temperature. To carry out the synthetic experiment, a synthetic T_{surf} was simulated using the prescribed mode for each combination of β_s and β_V between 0 and 1 and various climate (meteorological forcing) and vegetation cover (LAI) configurations. Each T_{surf} value was then forced as input after adding a random perturbation (white noise of 1 °C

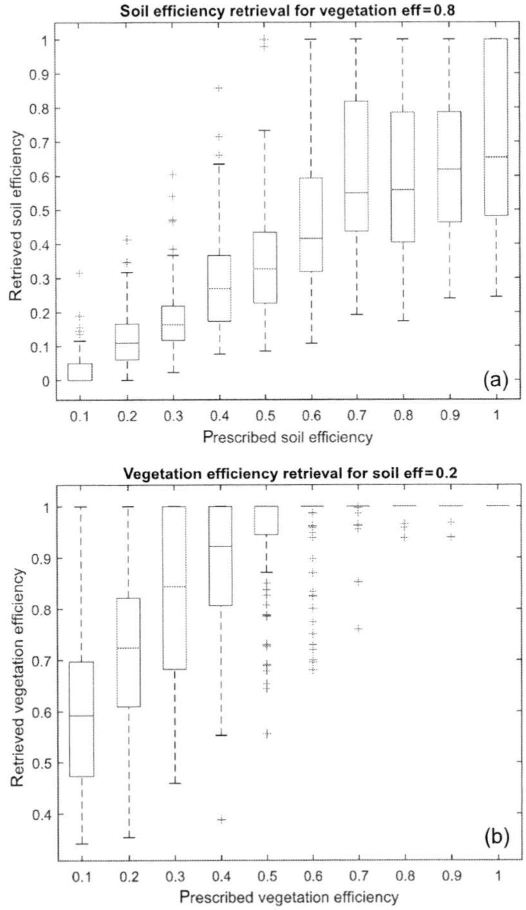

Figure 3. Retrieval statistics for conditions close to the model's assumption: β_s retrieval for $\beta_V = 0.8$ **(a)** and β_V retrieval for $\beta_s = 0.2$ **(b)**.

standard deviation) for the retrieval mode. Prescribed and retrieved stress levels were then compared: ideally, if the model is perfect, prescribed and retrieved efficiencies should match. Two situations were particularly looked at: one close to the model assumption (slight vegetation stress and low surface soil mositure) and one away from it (high vegetation stress and wet soil surface, such as after a small rain event). Results for a typical dry Mediterranean midday climate in spring (global solar radiation of 900 W m^{-2}, wind speed of 2 m s^{-1}, relative humidity of 40 %, and air temperature of 25 °C) and a LAI value close to 1.2 are illustrated on Figs. 2 to 4.

Figure 2 shows the total stress isolines for each combination of evaporation and transpiration efficiencies. Figure 3 shows the retrieved efficiencies when the conditions are close to initial assumptions used by the model either for the soil (dry soil with a prescribed efficiency of 0.2) or the vegeta-

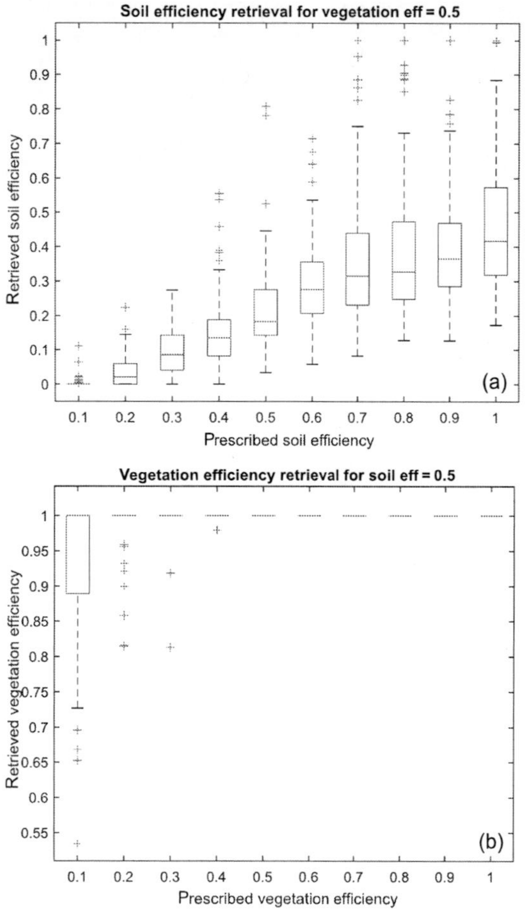

Figure 4. Same as Fig. 3 for conditions away from the model's assumptions: β_S retrieval for $\beta_V = 0.5$ **(a)** and β_V retrieval for $\beta_S = 0.5$ **(b)**.

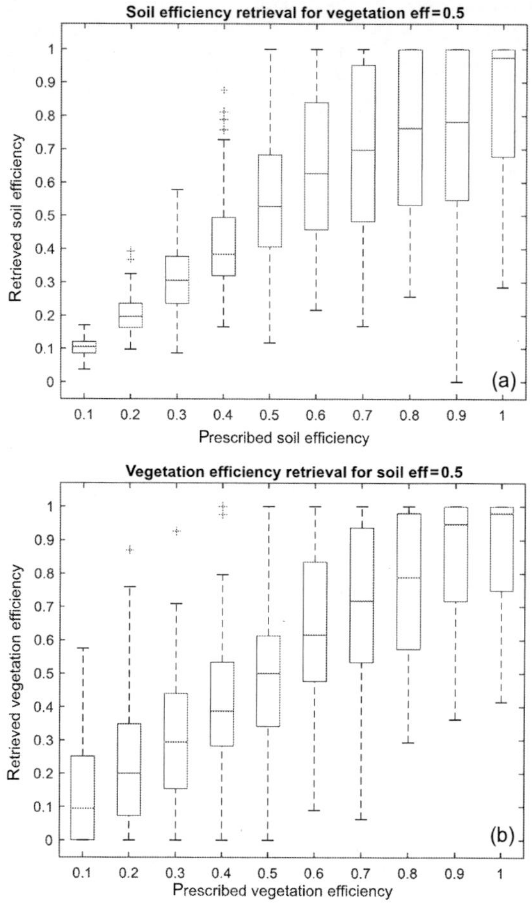

Figure 5. Same as Fig. 4 when both the surface temperature and the surface soil moisture are used as inputs instead of the sole surface temperature.

tion is transpiring at a level close to the potential rate (transpiration efficiency of 0.8). Figure 3 describes the retrieval of the soil efficiency when a fixed value of the transpiration of 0.8 is used (Fig. 3a) or the retrieved transpiration efficiencies when a fixed value of the soil evaporation efficiency of 0.2 is used (Fig. 3b), respectively. Figure 4 does the same for conditions away from the initial guess, that is a more pronounced vegetation stress (transpiration efficiency of 0.5, Fig. 4a) or a relatively wet soil (evaporation efficiency of 0.5, Fig. 4b). If total ET is always consistent between each prescribed and retrieved stress level combinations (Fig. 2), with prescribed and retrieved curve levels overlapping, it is not the case for the individual components. For instance, there is always a mismatch between retrieved (less stressed) and prescribed plant water stress levels, because the model always assumes as a first guess an unstressed canopy. This is true for the conditions close to the assumption (Fig. 3b) and even more pronounced for the conditions away from the assumption (Fig. 4b). Soil evaporation (and the associated relative efficiency) is less affected by this mismatch (Figs. 3a and 4a) and the retrieved soil evaporation efficiency follows the trend of the prescribed one, except maybe for the highest efficiency levels (retrieved efficiency reaches a plateau at around 0.6 if the prescribed transpiration efficiency is 0.8, and 0.4 if the transpiration efficiency is equal to 0.5).

3.2 Adding an additional constraint with respect to the soil evaporation efficiency

In order to improve the retrieval away from the initial assumptions, one decided to add another constraint to the retrieval process, the surface soil moisture, as it could be retrieved from remote sensing (i.e. Sentinel1 data).

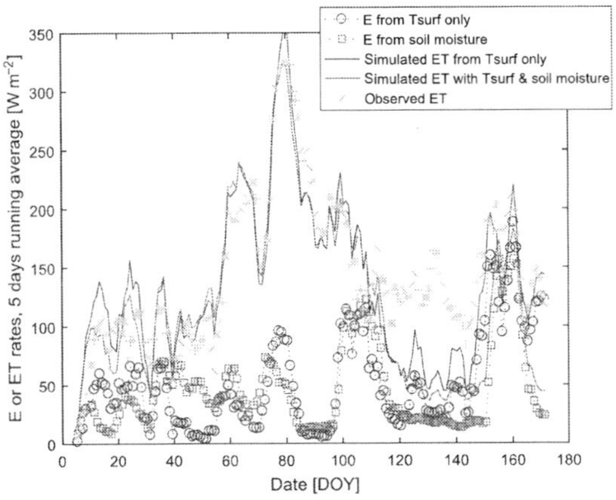

Figure 6. ET by both methods, observed ET, and soil E retrieved from the sole surface temperature or imposed from soil moisture level.

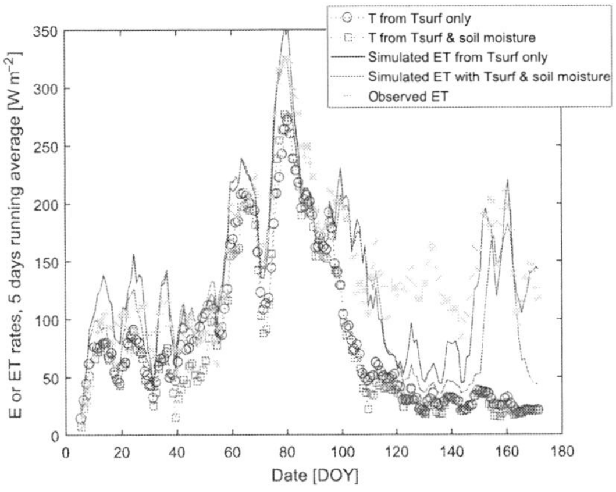

Figure 7. ET by both methods, observed ET, transpiration retrieved from the sole surface temperature or from combined surface temperature/soil moisture information.

Instead of using the initial guess of a fully transpiring canopy, we carry out the same experiment as in Sect. 3.1, but now evaporation is fixed by using a fixed value of the soil evaporation efficiency (as it could be derived from remote sensing). In order to account for the uncertainty related to the retrieval of soil moisture from radar data, a white noise with a 0.2 standard deviation is added to the efficiency. This value is taken from the order of magnitude of the total efficiency uncertainty from Boulet et al. (2015) at local scale and Saadi et al. (2018) for the irrigated perimeter scale. Results are presented in the Fig. 5 for conditions similar to Fig. 4 (conditions away from the model's original assumption), and show a large improvement for the vegetation efficiency retrieval (Fig. 4b).

4 Application to a real data obtained at a Mediterranean rainfed wheat site

In order to check the added value of surface soil moisture estimates to better characterize E, T and ET with real data, we did the same work for a rainfed wheat dataset (Boulet et al., 2015). The period of investigation spans an entire growing season (from emergence to harvest) and maximum LAI is around 2. ET is measured using an Eddy Covariance tower, while soil moisture at several depths is estimated from Thetaprobe soil moisture probes. An estimate of E (or its latent heat flux equivalent LE_s) is derived from the observed volumetric surface soil moisture is derived from Merlin et al. (2011):

$$LE_s = \left[0.5 - 0.5\cos\left(\pi\frac{\theta_{0\text{–}5\,\text{cm}}}{\theta_{\text{sat}}}\right)\right] LE_{sp} \quad (1)$$

where $\theta_{0\text{–}5\,\text{cm}}$ and θ_{sat} are the topsoil (0–5 cm) measured and the saturation soil moisture respectively.

Figures 6 and 7 show the retrieved evaporation and transpiration time series together with the total simulated observed ET. In order to smooth out some of the day-to-day fluctuations of ET due to varying incoming radiation, we present in the figures a moving average over five days of the daily fluxes. Midseason simulated ET is very close to the observations, and during that time E and T simulated using the sole temperature information or adding the soil moisture constraint are very similar. At the beginning (low LAI values) or the end (senescent vegetation) of the season the model reacts to the rainfall occurrence, but with an amplitude that is either too large (overestimation, such as for DOY 5–40) or too small (large underestimation at the end of the season, DOY 110–140). For the former period, there is certainly a signifiant uncertainty in LE_{spot} since both methods lead to similar results. For the latter period, the drop in green LAI induces a large decrease in transpiration, whereas some parts of the standing wheat are still transpiring.

The interesting result is that in fact the dual forcing does not improve the results (on the contrary, RMSE on LE at midday increases slightly from 57 to 67 $W\,m^{-2}$). When the sole surface temperature is used as input, the evaporation at the end of the season increases in order to decrease the surface temperature, which is not possible when soil moisture is imposed. It seems that when the model performs well, i.e. around maximum growth, both versions (with or without information on surface soil moisture) perform similarly, while the poorer performances of the model for early growth or during senescence (when the green LAI drops to zero while there is obviously some transpiration from the dry standing parts) explain the difficulty to take advantage of the additional constraint in that particular case, contrarily to the result shown in the synthetic experiment. There is therefore a need for further research to balance accurately the constraint of soil moisture and surface temperature for ensuring an increased robustness in ET estimates when adding additional information.

Data availability. For the data, one should contact the lead author. SPARSE model code available at http://tully.ups-tlse.fr/gilles.boulet/sparse (last access: 23 October 2018).

Competing interests. The authors declare that they have no conflict of interest.

Special issue statement. This article is part of the special issue "Earth Observation for Integrated Water and Basin Management: New possibilities and challenges for adaptation to a changing environment". It is a result of The Remote Sensing & Hydrology Symposium, Cordoba, Spain, 8–10 May 2018.

Acknowledgements. Funding from the CNES/TOSCA program for the PITEAS project is gratefully acknowledged. We also thank the International Joint Laboratory NAILA.

Edited by: William Kustas
Reviewed by: Juan Manuel Sánchez and three anonymous referees

References

Boulet, G., Mougenot, B., Lhomme, J.-P., Fanise, P., Lili-Chabaane, Z., Olioso, A., Bahir, M., Rivalland, V., Jarlan, L., Merlin, O., Coudert, B., Er-Raki, S., and Lagouarde, J.-P.: The SPARSE model for the prediction of water stress and evapotranspiration components from thermal infra-red data and its evaluation over irrigated and rainfed wheat, Hydrol. Earth Syst. Sci., 19, 4653–4672, https://doi.org/10.5194/hess-19-4653-2015, 2015.

Kustas, W. P. and Norman, J. M.: Evaluation of soil and vegetation heat flux predictions using a simple two-source model with radiometric temperatures for partial canopy cover, Agr. Forest Meteorol., 94, 13–29, https://doi.org/10.1016/s0168-1923(99)00005-2, 1999.

Merlin, O., Al Bitar, A., Rivalland, V., Beziat, P., Ceschia, E., and Dedieu, G.: An Analytical Model of Evaporation Efficiency for Unsaturated Soil Surfaces with an Arbitrary Thickness, J. Appl. Meteorol. Climatol., 50, 457–471, https://doi.org/10.1175/2010jamc2418.1, 2011.

Saadi, S., Boulet, G., Bahir, M., Brut, A., Delogu, É., Fanise, P., Mougenot, B., Simonneaux, V., and Lili Chabaane, Z.: Assessment of actual evapotranspiration over a semiarid heterogeneous land surface by means of coupled low-resolution remote sensing data with an energy balance model: comparison to extra-large aperture scintillometer measurements, Hydrol. Earth Syst. Sci., 22, 2187–2209, https://doi.org/10.5194/hess-22-2187-2018, 2018.

Schmugge, T. J.: Effect of texture on microwave emission from soils, IEEE T. Geosci. Remote, 18, 353–361, 1980.

Proc. IAHS, 380, 23–28, 2018
https://doi.org/10.5194/piahs-380-23-2018

Open Access PIAHS

Use of canopy coefficients obtained from satellite data to estimate evapotranspiration over high mountain Mediterranean watersheds

Elisabet Carpintero[1], **María P. González-Dugo**[1], **Jorge Jódar**[2], **and Sergio Martos-Rosillo**[3]

[1]IFAPA-Centro Alameda del Obispo, Apdo. 3092 14080 Cordoba, Spain
[2]Groundwater Hydrology Group, Dept. Civil and Environmental Eng., Technical University of Catalonia (UPC), Hydromodel Host S.L. and Aquageo Proyectos S.L., Spain
[3]Geological Survey of Spain (IGME), Granada, Spain

Correspondence: Elisabet Carpintero (elisabet.carpintero@juntadeandalucia.com)

Received: 23 April 2018 – Revised: 5 September 2018 – Accepted: 14 September 2018 – Published: 18 December 2018

Abstract. This work explores the dynamics of the water consumed by the vegetation in two Mediterranean watersheds of Sierra Nevada Mountains (Southern Spain). This region has experienced an increase in the demand of water in the last years due to the growth of irrigated areas, and a new water resources plan is required. The evapotranspiration (ET) of irrigated horticultural crops and natural communities were monitored for the hydrological years 2013/14 and 2014/15, using an approach based on the concept of reference evapotranspiration (ET_0) and canopy coefficients derived from the computation of vegetation indices (VIs), which we will call the VI-ET_0 approach. A set of Landsat-8 and MODIS images has been used as remote input data. The results were used for the spatial analysis of water consumption in terms of the main land cover types in the area. The annual runoff obtained with a simple surface water balance, using the ET values estimated by the VI-ET_0 approach, was comparable to that obtained by the HBV (Hydrologiska Byrans Vattenbalansavdelning) model, a precipitation-runoff generation model that reproduces the observed river discharge at the outlet of the watershed.

1 Introduction

The high mountain areas of Sierra Nevada (Southern Spain) have a great ecological value due to the presence of endemic ecosystems together with landscapes and cultural heritage, being protected by the status of Biosphere Reserve (1986) and National Park (1999). However, a considerable increase of irrigated areas has been observed in the region over the last few years, resulting in a rise in groundwater consumption, which in turns has an impact on the functioning regime of rivers and the water quality of the basins runoff. This led to the river basin authority to conduct a pilot study in order to improve the current knowledge of the hydrological behaviour of the high mountain area of Sierra Nevada. In this context, the dynamics of water consumption in the basin, including both irrigated horticultural crops and natural vegetation communities, should be continuously monitored to guarantee a sustainable management of water resources in this area.

Remote sensors provide distributed information about the status of vegetation. When this information is combined with meteorological data and integrated into models for estimating evapotranspiration (ET), it becomes a tool that allows a regular monitoring of water use at a regional scale. An extended approach in agricultural applications to estimate crop water requirements is the one proposed by Allen et al. (1998) FAO56 guidelines. It determines the actual ET using the concepts of crop coefficient and reference ET (ET_0). There is an approach that combines this method with the spectral reflectance of vegetated surfaces, which is provided by remote sensors in the visible and near infrared regions through vegetation indices (VIs), hereunder called the VI-ET_0 approach (Campos et al., 2016; González-Dugo et al., 2009). As a result, the potential transpiration of the plant is determined in terms of a spatially-distributed basal crop coefficient.

The objective of this work is to estimate, on a regular and spatially distributed basis, the water consumption of the vegetation in two neighbouring high mountain watersheds of Sierra Nevada, for the hydrological years 2013/14 and 2014/15 by applying the VI-ET_o approach. A water resources evaluation in the watersheds is conducted combining the VI-ET_o approach with both, the results obtained from the hydrological HBV (Hydrologiska Byrans Vattenbalansavdelning) model (Bergström et al., 1976) and river runoff data measured in situ. Finally, the pattern of water consumption in terms of the main land cover types in the area is analyzed.

2 Study area

The study area is composed by the two adjacent watersheds of Bérchules and Mecina, located on the southern slopes of Sierra Nevada. The former one is located to the west and the latter to the east, with the intervening north-south water divide. Both watersheds are similar in size, shape, orientation, vegetation, soil and geological bedrock. The Bérchules watershed has an area of 68 km^2, with an altitude which rises from 980 m a.s.l. at the outlet to about 2910 m at the highest point. The Mecina watershed is slightly smaller than Bérchules, it has an area of 55 km^2 and presents an elevation ranging between 1042 and 2618 m a.s.l. In terms of land cover types, the study area is mainly characterized by natural vegetation of grassland, scrubs, conifers and oaks, followed by irrigated horticultural crops and rainfed fruit trees.

3 Methods and Materials

3.1 Remote sensing-based model

The VI-ET_o approach is based on the concepts of ET_o and the VIs derived crop/canopy coefficient (K_c). The former term, which takes into account the atmospheric demand, is calculated in this work with the Hargreaves equation (Hargreaves et al., 1985). The latter one, K_c, considers the effect of a specific canopy in relation to a given reference grass surface and was obtained using the dual method in the form popularized by FAO (Allen et al., 1998). It separates canopy transpiration from soil surface evaporation and compute evapotranspiration as follows:

$$\mathrm{ET} = (K_{cb} K_s + K_e)\mathrm{ET_o} \tag{1}$$

being $K_c = K_{cb} K_s + K_e$, where K_{cb} is the basal canopy coefficient and determines the vegetation transpiration, K_s is the water stress coefficient, that quantifies the reduction in transpiration due to soil water deficits, and K_e is the soil evaporation coefficient.

In this work, K_{cb} is derived from the canopy spectral response, provided by a set of satellite images. Numerous empirical relationships between VI-K_{cb} have been proposed for herbaceous and tree crops, (Consoli and Barbagallo, 2012; Odi-Lara et al., 2016) and natural canopies (Campos et al., 2013). The equation proposed by González-Dugo et al. (2009) to compute the basal crop coefficient from the SAVI index (Soil Adjusted Vegetation Index; Huete, 1988) is used here, and it can be expressed as:

$$K_{cb} = \begin{cases} \dfrac{K_{cb,max}}{f_{cb,max}} \left(\dfrac{\mathrm{SAVI} - \mathrm{SAVI_{min}}}{\mathrm{SAVI_{max}} - \mathrm{SAVI_{min}}} \right); & f_c < f_{c,max} \\ K_{cb,max}; & f_c \geq f_{c,max} \end{cases} \tag{2}$$

where the subscripts max and min refer to the values of SAVI for a full ground vegetation coverage and bare soil, respectively. The ratio $(\mathrm{SAVI} - \mathrm{SAVI_{min}})/(\mathrm{SAVI_{max}} - \mathrm{SAVI_{min}})$ is an estimation of f_c, and $f_{c,max}$ is the canopy ground-cover fraction (f_c) at which K_{cb} is at its maximum ($K_{cb,max}$). $K_{cb,max}$ is a canopy dependant parameter, with tabulated values taken from Allen et al. (1998). This equation has been validated with satisfactory results over a wide range of annual crops and fruit tree crops over the Mediterranean conditions of Southern Spain (González-Dugo et al., 2013; Mateos et al., 2013).

The soil evaporation coefficient K_e is obtained as:

$$K_e = K_r \left(K_{c,max} - K_{cb} \right) \tag{3}$$

Where K_r is a dimensionless evaporation reduction coefficient that depends on topsoil water depletion and $K_{c,max}$ is the maximum value of K_c following a rainfall or irrigation event. The value of K_e cannot be higher than the product ($f_{ew} \times K_{c,max}$), where f_{ew} is the fraction of the soil surface that is both exposed and wetted (Allen et al., 1998).

The stress coefficient K_s is obtained on a daily basis by computing a soil water balance in the root zone. The daily root zone water deficit can be expressed as:

$$\mathrm{RZWD}_i = \mathrm{RZWD}_{i-1} + \mathrm{ET}_{i-1} - R_i \tag{4}$$

where the subscript i indicates a given day, RZWD_i and RZWD_{i-1} are the root zone water deficit on days i and the previous one respectively, and R is rainfall or irrigation. When the root zone water content is at field capacity, the soil water content is maximum ($\mathrm{RZWD} = 0$), while this value is minimum at the wilting point. The depth of water between both extremes is named root zone water holding capacity (RZWHC). Equation (5) calculates the water stress coefficient K_s based on the relative root zone water deficit.

$$K_s = \begin{cases} \dfrac{\mathrm{RZWHC} - \mathrm{RZWD}_i}{(1-p)\mathrm{RZWHC}}; & \mathrm{RZWD}_i < (1-p)\mathrm{RZWHC} \\ 1; & \mathrm{RZWD}_i \geq (1-p)\,\mathrm{RZWHC} \end{cases} \tag{5}$$

where p is the fraction of the RZWHC below which transpiration is reduced as a consequence of water deficit.

This simplified soil water balance does not take into account the lateral movement of the water which, under the

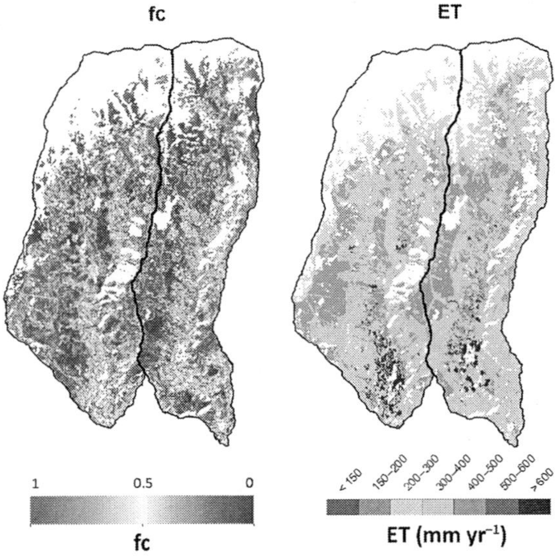

Figure 1. Distributed average values of f_c and annual estimated ET values for the hydrological year 2013/14.

conditions irrigated or little slope areas with scarce runoff, has a reduced impact. However, in areas with high steepness and low ground cover vegetation, as some part of this watershed, this simplification may increase the modelling error of ET. The limitations of the model under different conditions of steepness and ground fraction cover require further analysis.

The soil water balance computation was initiated in August 2013. The root zone water content was taken at field capacity on irrigated horticultural crops, while for the remaining vegetation types, the soil was considered dry and the root zone water deficit was assumed to be equal to the soil water holding capacity. It was established a standard irrigation schedule for the irrigated crops, based on common practices in the area. Irrigations every two days were applied by refilling the soil profile until reaching the field capacity during the irrigation season, from mid-May to mid-October. Selected values of parameters p and $K_{c,max}$ for each land cover are described in Jódar et al. (2018).

3.2 Description of input data

A set of 26 cloud-free Landsat-8 images between August 2013 and October 2015 was selected and atmospherically corrected to obtain surface reflectance values. The optical bands correction was performed with the radiative transfer model Modtran4 (Berk et al., 1989), and using standardized parameters to characterize the state of the atmosphere at the acquisition time. Moreover, four MODIS images (product MOD09GQ) were added to complete the input data series. During the winter season, the low sun elevation angles produced that the steepest areas were not illuminated at sensor overpass time. These areas have been masked out during the image processing and their subsequent use. Daily meteorological data were provided by a network of automatic weather stations, including variables of rainfall, solar radiation, air temperature and humidity, and wind speed. The point data were spatially interpolated to obtain distributed information with the hydrological model WiMMed (Herrero et al., 2011) using the inverse distance IDW method, and taking into account the altitudinal factor.

The vegetation types have been obtained from the Information and Soil Uses System of Spain (SIOSE), which have been grouped into different classes with similar spectral response. For refining the identification of irrigated horticultural crops areas, satellite images and data provided by Environment and Water Agency of Andalusia (AMAYA) have also been used. Soil parameters (texture, water content at field capacity and wilting point, and soil depth) are provided by regional soil maps produced by the Agricultural Department of Andalusia at 250 m of spatial resolution (CAPMA, 2008).

3.3 Evaluation of water resources

Daily ET values were aggregated at annual and watershed scales, and a simple surface water balance was computed to estimate annual watershed runoff (Q_{runoff}) as follows:

$$Q_{runoff} = P - ET - I \tag{6}$$

where P is rainfall and I is rainfall interception by vegetation, which is assumed as 8 % of rainfall according to Polo et al. (2013). The annual runoff values obtained with the VI-ET_0 approach were compared with both, the observed runoff values at the Narila gauging station (located at the outlet of the Bérchules watershed) and the runoff values simulated by the HBV model. HBV is a conceptual rainfall-runoff model for catchment hydrology modeling. It is based on the general water balance and provides daily basin discharge by subjecting daily precipitation to four conceptual modules: a snow routine, a soil moisture routine, a response routine and a routing routine (Bergström et al., 1976).

4 Results

4.1 The VI-ET_0 model

One of the main variables required by this approach to compute the water consumption of the different canopies is the evolution of their vegetation ground cover fraction (f_c), which keeps a linear relationship with the values of SAVI derived from the satellite images. Figure 1 shows the average f_c values for both watersheds and the first hydrological year (2013/14). The lowest values can be found in the upper parts of the watersheds, corresponding to high mountain grasslands and scrubs. The conifers areas, located at the

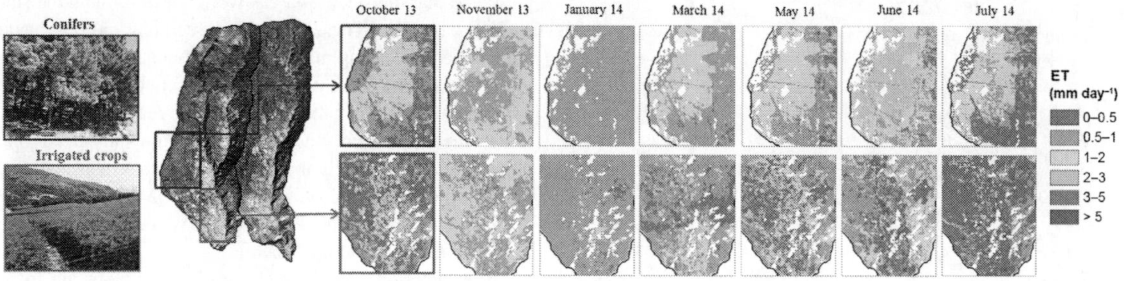

Figure 2. Evolution of the distributed values of daily ET over areas with two representative vegetation: conifers (dark-blue line) and irrigated horticultural crops (line red) for Bérchules watershed.

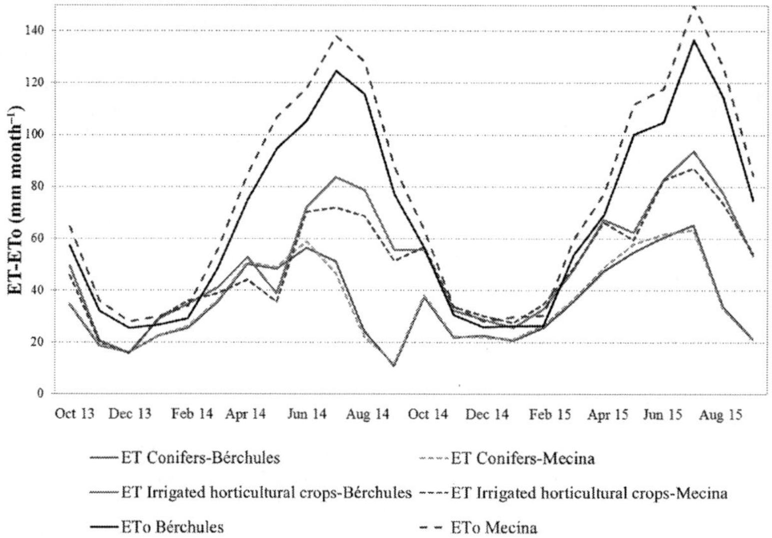

Figure 3. Monthly ET_0 and estimated ET values of two representative canopies: conifers and irrigated horticultural crops for the Bérchules and the Mecina watersheds.

western sunny hillside of each watershed, present the higher and more stable f_c values along the year. The irrigated horticultural crops are predominant in the South, with significant changes during the year.

Spatially distributed values of canopy water consumption were obtained at 30 m spatial resolution and on a daily time step, which were later aggregated to monthly and annual time steps. In Fig. 1, annual ET values are also shown for the hydrological year 2013/14. Average values estimated for the two hydrological years (2013/14 and 2014/15) were 353.5 mm in Bérchules and 352.2 mm in Mecina. These numbers support the assumption of a similar hydrological behaviour of both watersheds.

The annual runoff values obtained by the VI-ET_0 approach were 81.9 and 83.2 mm (year 2013/14 and 2014/15 respectively) for the Bérchules watershed, while for Mecina are 45.9 and 58.1 mm. There is a reasonable agreement between the simulated data obtained by both models (HBV and VI-ET_0 approach), with a RMSE value of 4.4 and 5.5 mm for Bérchules and Mecina, respectively. Moreover, the RMSE value obtained with the VI-ET_0 approach when comparing the observed and simulated data for the Bérchules watershed is of 4.5 mm. A more detailed description on HBV model application and comparison results discussion can be found in Jódar et al. (2018).

4.2 ET patterns of the main land cover types

Distributed daily ET values over areas with two representative vegetation types, conifers and irrigated horticultural crops, for some months of the year 2013/14 are shown in Fig. 2. These land covers present a particular interest as their surfaces in the area have experienced a significant increase in the last decades. Jiménez-Olivenza et al. (2015) quantified a 22.5 % of increase in the conifer area of Sierra Nevada since 1956 to 2011 and, even when a reduction in agricultural area around 12 % was reported by these authors, the river basin authority has observed in more recent years a clear rise of intensive horticulture. Both changes have significant implications on water consumption at the watershed scale. High daily water consumptions ($> 5\,\mathrm{mm\,day}^{-1}$) are estimated over the irrigated areas during the summer months (crop growth and irrigation season) as can be observed in Fig. 2. Likewise, conifers areas can be well identified with high ET values in June (2–3 $\mathrm{mm\,day}^{-1}$). The temporal evolution of monthly ET_0 and estimated actual ET of conifers and irrigated horticultural crops covers are depicted in Fig. 3.

Conifers present intra-annual changes in ET coupled with variations in the evaporative demand of the atmosphere, while maintaining quite regular vegetation coverage along the year. The ET evolution of irrigated horticultural crops follows a seasonal pattern, with a main annual peak around the middle of the summer (between 69 and 93 $\mathrm{mm\,month}^{-1}$), with crops starting their growing cycle in May and presenting quite low values during the end of autumn and winter.

In quantitative terms, the irrigated agriculture presents the highest values of water consumption (604.1 $\mathrm{mm\,year}^{-1}$ on average). However, the area of this land use represents about 4 % of the total area in both cases, leading to a small contribution (6–7 % on average) to the watershed total water use. On the other hand, conifer areas present a proportionally high contribution of 18.5 % to total ET with a higher coverage than horticultural crops (around 15 %) but lower annual consumption (422 $\mathrm{mm\,year}^{-1}$ on average). Considering that before 1956 the conifer area was almost null and the soil was occupied by grassland and scrubs (Jiménez-Olivenza et al., 2015) with a lower annual consumption of water, these results highlight the importance of considering not only agricultural uses when attending to the water management of these watersheds.

5 Conclusions

The VI-ET_0 approach has been used to estimate the spatio-temporal variation of ET in the Bérchules and the Mecina watersheds, using the vegetation indices derived from satellite images and other auxiliary information.

The analysis of water use patterns of the different land covers, in particular conifer and intensive horticulture, suggest that, even when the consumption of agricultural area is, on average, higher, the contribution of conifer and other forest uses to the total ET of the watershed is significant. The land use changes of these natural canopies in the last 60 years have had a quantitatively higher impact on total water consumption in the watershed than current increases in agricultural areas. A continuous monitoring of land uses changes and evapotranspiration using this procedure could be useful to assist future decisions leading to guarantee a sustainable management of water resources of this area.

Data availability. This work is part of the project "Estudio de las cubiertas vegetales de la cabecera de los rios grande de bérchules y mecina aplicando técnicas de teledetección", commissioned by D. G. de Planificación y Gestión del Dominio Público Hidráulico of Junta de Andalucía. Please, contact the authors if you are interested in this research data.

Author contributions. The authors EC and MPGD designed the application of the VI-ETo approach in the study area. EC carried out the application of the remote sensing-based model. JJ simulated the annual runoff values with the HBV model. SMR provided the observed data at the outlet of the watershed. All the authors contributed in the discussion of the results. Finally, EC and MPGD prepared the manuscript with the contributions of the rest of the authors.

Competing interests. The authors declare that they have no conflict of interest.

Special issue statement. This article is part of the special issue "Earth Observation for Integrated Water and Basin Management: New possibilities and challenges for adaptation to a changing environment". It is a result of The Remote Sensing & Hydrology Symposium, Cordoba, Spain, 8–10 May 2018.

Acknowledgements. The authors would like to recognize the funding supported by FPU program – University Teachers Training – from Department of Education, Culture and Sport. Additional support was provided by the National Agricultural and Food Research and Technology Institute, INIA, and European Fund for Regional Development 2014-2020 (Operative programme of smart growth) through project RTA2014-00063-C04-02. The project ESTUDIO DE LAS CUBIERTAS VEGETALES DE LA CABECERA DE LOS RIOS GRANDE DE BÉRCHULES Y MECINA APLICANDO TÉCNICAS DE TELEDETECCIÓN (PP.PEI.IDF201401.9) funded by European Fund for Regional Development (Operative programme of Andalucia 2007–2013) also provided support.

Edited by: Guido D. Urso
Reviewed by: William Kustas and two anonymous referees

References

Allen, R. G., Pereira, L. S., Raes, D., and Smith, M.: Crop Evapotranspiration: Guidelines for Computing Crop Requirements, Irrigation and Drainage Paper No. 56, FAO, Roma (Italia), 1998.

Bergström, S.: Development and Application of a Conceptual Runoff Model For Scandinavian Catchments, SMHI, Report No. RHO 7, Norrköping, 1976.

Berk, A., Bernstein, L. S., and Robertson, D. C.: MODTRAN: a moderate resolution model for LOWTRAN7, GL-TR -89-0122, Air Force Geophys. Lab., Hanscom AFB, MA, 38 pp., 1989.

Campos, I., Villondre, J., Carrara, A., and Calera, A.: Remote sensing-based soil water balance to estimate Mediterranean holm oak savanna (dehesa) evapotranspiration under water stress conditions, J. Hydrol., 494, 1–9, 2013.

Campos, I., Balbontín, C., González-Piqueras, J., González-Dugo, M. P., Neale, C. M. U., and Calera, A.: Combining a water balance model with evapotranspiration measurements to estimate total available soil water in irrigated and rainfed vineyards, Agr. Water Manage., 165, 141–152, 2016.

Consejería de Agricultura y Pesca, Junta de Andalucía (CAPMA): Sistema de Inferencia Espacial de Propiedades Físico-Químicas e Hidráulicas de los Suelos de Andalucía (Informe Final), 2008.

Consoli, S. and Barbagallo, S.: Estimating water requirements of an irrigated mediterranean vineyard using a satellite-based approach, J. Irrig. Drain. E., 138, 896–904, 2012.

González-Dugo, M. P., Neale, C. M. U., Mateos, L., Kustas, W. P., Prueger, J. H., Anderson, M. C., and Li, F.: A comparison of operational remote sensing-based models for estimating crop evapotranspiration, Agr. Forest Meteorol., 149, 1843–1853, 2009.

González-Dugo, M. P., Escuin, S., Mateos, L., Cano, F., Cifuentes, V., Padilla, F. L. M., Tirado, J. L., Oyonarte, N., and Fernández, P.: Monitoring evapotranspiration of irrigated crops using crop coefficients derived from time series of satellite images. II. Application on basin scale, Agr. Water Manage., 125, 92–104, 2013.

Hargreaves, G. L., Hargreaves, G. H., and Riley, J. P.: Irrigation water requirements for Senegal river basin, J. Irrig. Drain. E., 111, 265–275, 1985.

Herrero, J., Millares, A., Aguilar, C., Díaz, A., Polo, M. J., and Losada, M. A.: WiMMed. Base Teórica, Grupo de Dinámica fluvial e Hidrología de la Universidad de Córdoba y Grupo de Dinámica de Flujos Ambientales de la Universidad de Granada, Spain, 2011.

Huete, A. R.: A soil-adjusted vegetation index (SAVI), Remote Sens. Environ., 25, 295–309, 1988.

Jiménez-Olivenza, Y., Porcel-Rodríguez, L., Caballero-Calvo, A., and Bonet, F. J.: Evolución de los usos del suelo en Sierra Nevada en los últimos 50 años y cambios del paisaje, La huella del Cambio Global en Sierra Nevada: Retos para la conservación, Consejería de Medio Ambiente y Ordenación del Territorio, Junta de Andalucía, 54–56, 2015.

Jódar, J., Carpintero, E., Martos-Rosillo, S., Ruiz-Constán, A., Marín-Lechado, C., Cabrera-Arrabal, J. A., Navarrete-Mazariegos, E., Gonzáles-Ramón, A., Lambán, L. J., Herrera, C., and González-Dugo, M. P.: Combination of lumped hydrological and remote-sensing models to evaluate water resources in a semi-arid high altitude ungauged watersehd of Sierra Nevada (Southern Spain), Sci. Total Environ., 625, 285–300, 2018.

Mateos, L., González-Dugo, M. P., Testi, L., and Villalobos, F. J.: Monitoring evapotranspiration of irrigated crops using crop coefficients derived from time series of satellite images. I. Method validation, Agr. Water Manage., 125, 81–91, 2013.

Odi-Lara, M., Campos, I, Neale, C. M. U., Ortega-Farías, S., Poblete-Echeverría, C., Balbontín, C., and Calera, A.: Estimating Evapotranspiration of an Apple Orchard Using a Remote Sensing-Based Soil Water Balance, Remote Sens., 8, 253, https://doi.org/10.3390/rs8030253, 2016.

Polo, M. J., Aguilar, C., Díaz-Gutiérrez, A., and González-Dugo, M. P.: Cuantificando la interceptación en cuencas heterogéneas: la cuenca del río Guadalfeo (Granada), Interceptación de la lluvia por la vegetación en España, Instituto Euromediterráneo del agua, 297–317, 2013.

Proc. IAHS, 380, 29–35, 2018
https://doi.org/10.5194/piahs-380-29-2018

Open Access PIAHS

Remote sensing-based soil water balance for irrigation water accounting at the Spanish Iberian Peninsula

Jesús Garrido-Rubio[1], **Alfonso Calera Belmonte**[1], **Lorena Fraile Enguita**[1], **Irene Arellano Alcázar**[1], **Mario Belmonte Mancebo**[1], **Isidro Campos Rodríguez**[1], **and Raquel Bravo Rubio**[2]

[1]Remote Sensing & GIS Group, Institute for Regional Development (IDR), University of Castilla-La Mancha (UCLM), Albacete, 02071, Spain
[2]Subdirección General de Regadíos y Economía del Agua (SGRyEA), Ministerio de Agricultura y Pesca, Alimentación y Medio Ambiente (MAPAMA), Madrid, 28005, Spain

Correspondence: Jesús Garrido-Rubio (jesus.garrido@uclm.es)

Received: 17 April 2018 – Revised: 16 July 2018 – Accepted: 14 August 2018 – Published: 18 December 2018

Abstract. Temporal series maps of irrigated areas, and the corresponding irrigation water requirements based on remote sensing, is a recognized tool contributing to water governance at different scales, from water user associations to whole river basin districts. These thematic cartographies offer a first estimation of the crop irrigation requirements, and a biophysical based approach of the temporal and spatial distribution of the crop water use in the cultivated areas. This work describes the operational application of these methodologies, providing valuable information for water governance and management purposes. The basic products obtained in the whole Spanish part of the Iberian Peninsula during the period 2014–2017 were: (i) annual maps of irrigated crops based on time series of multispectral satellite imagery; and (ii) the direct remote sensing-based water accounting, by quantifying agricultural water flows (e.g. rainfall, irrigation, evapotranspiration, drainage and recharge), through a remote sensing-based soil water balance. Hence this paper provides a remote sensing based water accounting approach, which relies on dense time series of multispectral imagery acquired by the multisensor constellation arranged by Landsat 8 and Sentinel-2 satellites, jointly with meteorological data and agronomic knowledge. Then, based on these purpose and approach, annual and monthly maps of net irrigation water requirements have been elaborated at the most practical spatial and temporal scales for water governance purposes over big areas such river basin districts. This work summarizes the methodologies used and discuss the technical and non-technical feasibility of the proposed approach.

1 Introduction

The monitoring of irrigated areas and the accounting of their water use are essential instruments for European policies, in order to promote an efficient and sustainable use of water, according to the Water Framework Directive, WFD (2000/60/EC, 2000), and transposed into each country member legislation.

Monitoring of crops and their water requirements are possible through data derived from the Earth Observation, EO, combined with meteorological information. The Copernicus program of the European Union, through the European Space Agency (ESA), operates a constellation of Sentinel satellites, which provide data from the Earth's surface at a spatial resolution and a temporal frequency unprecedented in the history of remote sensing. Consequently the Copernicus programs exhibit a great potential to perform the monitoring required by the WFD.

Aware of the great socio-economic and environmental importance of irrigation in Spain, the Ministry of Agriculture and Fisheries, Food and Environment (MAPAMA) pushes actions to improve the efficiency of water use on irrigation. Among other, through the Deputy Directorate General for Irrigation and Water Economy (SGRyEA), it makes available freely to users the information collected by the network of agrometeorological automatic stations, known as the SIAR network (Agroclimatic Information System for Irrigation). A step further is the SPIDER-SIAR 2014–2017 project, which joins the potential provided by the EO programs for the spatial and temporal description of the vegetation cover, to the agrometeorological information provided by the SIAR network, all for the knowledge of irrigated areas and their water use in the area of mainland Spain.

2 Objective, products and stakeholders

The main objectives of the SPIDER-SIAR project are, (1) improving water efficiency, and (2) better managing water resources by public authorities through mapping irrigated areas, estimating surface cover, and determining their net irrigation water requirements (NIWR) in the mainland Spain (2014–2017 period). Spanish main irrigation features are summarized as a surface that covers around 3.7 Mha, with wheat, barley, maize, vineyard, olive, fruit trees, and citric as main crops, and mainly irrigated by drip and sprinkle systems (around 75 %). Regarding such spatial extent, SPIDER-SIAR mainly uses a time series of satellite images of the constellation Sentinel-2A (S2A), Sentinel-2B (S2B) and Landsat 8 (L8), together with agrometeorological information of the SIAR network and ancillary information.

Methodology and materials used allow obtaining results from a minimum area of 0.1 ha, based on a daily temporal scale, and able to monitor all typologies of agricultural surfaces. These primary results are suitable to aggregate into other scales of spatial and temporal interest.

The beneficiaries of the project are at the same time water users and water managers, who can share the data at their different administrative scales, like irrigation communities, river basin authorities, regions, nation-wide and European levels, which deal with the planning of water resources in the medium and long term, through the execution and monitoring of river basin management plans (RBMPs). All the products can be accessed freely through a WebGIS platform called SPIDER-SIAR. Table 1 details the products and relates them to their potential users.

3 Materials and methodology

3.1 Time series of satellite images

The images used are those acquired, in the area of the mainland Spain, along the period 2014 to 2017, by the multispectral sensors on board the S2A, S2B and L8 platforms. The European Space Agency (ESA) and the National Aeronautics and Space Administration (NASA) from USA operate these platforms and distribute freely the images. Once those high cloud cover images have been discarded, a total number of 13 006 images were processed, of which 2437 scenes correspond to L8, 7957 granules to S2A, and 2612 granules to S2B. These images are then converted into the three basic SPIDER-SIAR products: (i) Vegetation Index (VI) time series of NDVI (Rouse et al., 1973); (ii) an appropriate colour composition images (RGB) to monitor irrigated agriculture; and iii) the biophysical crop parameter Kcb. These time series products (unless RGB) are atmospherically corrected by absolute normalization (Campos et al., 2011). Additionally, cloudy and shadows pixels were removed using L8 Quality Band and Sen2Cor algorithms for S2A and S2B. Consequently, a virtual multisensor constellation was built to provide time series of NDVI able to describe accurately the temporal evolution of the vegetation cover.

3.2 Thematic maps of irrigated land use

For mapping the irrigated crops, the approach uses the vegetation temporal evolution pattern as a basic criterion, which is well described by the temporal trajectory of the NDVI values of each pixel. Moreover, to guide spatially the classification analysis, the approach integrates multiple ancillary official sources, especially the layer of SIGPAC enclosures (Spanish national layer to attend Common Agricultural Policy System Information).

The methodology assumes that: (1) the water supply produces differences in size and duration of irrigated vegetation cover with respect to the one not; (2) time series of VI images are able to record these differences; and hence (3), decisions-tree rules can be applied to segment between irrigated and non-irrigated crops. Consequently, the knowledge about the temporal evolution of the different crop canopy cover is the essential instrument in the irrigated crops identification. It is necessary to highlight that crops that present a similar evolution both in their phenology and in soil fraction cover, will show a similar evolution in their NDVI time trajectory. In those crops that exhibits a high proportion of bare soil, especially woody crops, and in which the irrigation reaches only a supplementary value, the NDVI time series among irrigated and non-irrigated exhibits small differences.

SPIDER-SIAR annual thematic maps of irrigated areas present 9 classes in which the irrigated crops are grouped: four of these classes group the annual herbaceous crops depending on the season when the crop growing cycle occurs (spring, summer, spring-summer, and autumn-winter); four more classes are used to group woody crops (vines, olives, citrus and deciduous fruit trees); and finally, one class for crops developed under greenhouses.

3.3 Thematic maps of Net Irrigation Water Requirements (NIWR)

Thematic maps of net irrigation water requirements (NIWR) of those irrigated areas (excluding greenhouses), are estimated by carrying out a water balance in the root soil layer. This soil water balance (SWB) follows the dual coefficient approach described in the FAO56 manual (Allen et al., 1998). For each pixel, we use the time series of NDVI values to estimate the value of the Kcb, what it is called "*reflectance-based crop coefficient approach*" (Neale et al., 1989). We se-

Table 1. SPIDER-SIAR products and scope of its services regarding different water user types.

Branch	Products offered	Scope of services
Remote sensing images processed from L8, S2A and S2B satellites	Time series of multispectral EO products: VI (NDVI); Colour images (RGB); and Basal crop coefficient (Kcb)	Crop monitoring at plot scale level in near real time by irrigation farmers or agricultural consultant
Irrigated crop classification	Annual thematic maps of irrigated crop classification	Crop monitoring at different work scales from water users associations to river basin water managers
Soil water balance. Net irrigation water requirements estimation	Annual and monthly thematic maps of net irrigation water requirements	Crop irrigation monitoring at different working scales from irrigation farmers to water user managers at water user association or river basin scales
Information sharing via WebGIS services	Sharing platform SPIDERwebGIS®: group SPIDER-SIAR	http://maps.spiderwebgis.org/login/?custom=spider-siar (last access: 13 September 2018)

lect a unique crop-independent NDVI-Kcb relationship using Eq. (1) (Campos et al., 2010), according with the real time and operational purpose at large scales (Calera et al., 2017), like other projects are dealing with, i.e. the case of California State at USA (Melton et al., 2012).

$$\mathrm{Kcb} = 1.44 \cdot (\mathrm{NDVI}) - 0.10 \quad (1)$$

Furthermore, the use of a single equation provided some benefits for the operational monitoring of NIWR. For applications at large scales, our actual knowledge of the crop biophysical variables such as LAI and canopy resistance must be considered unknown. Note that the relationship between LAI and VI is crop dependent and strongly affected by the canopy architecture (Campos et al., 2017; Anderson et al., 2004) while the available information about land uses is restricted to general crop types (see Sect. 3.2) and difficulty improved. In addition, although the proposed relationship was originally developed for irrigated-row structured vineyard, the parameterization of the function does not differ from other relationships obtained for herbaceous crops like corn (Bausch and Neale, 1987; Gonzalez-Piqueras et al., 2004), or biomass sorghum (López-Urrea et al., 2016). In the same way, the used equation had been evaluated in bush-vines in Spain (Campos et al., 2016), vineyards in Australia (Hornbuckle, 2014), natural vegetation (Campos et al., 2013), and table grapes trained in overhead trellises system (Balbontín et al., 2017).

In parallel, and in order to obtain NIWR, daily data from meteorological stations of the dense SIAR network, which placed strategically on the irrigated areas, enable to obtain reference evapotranspiration and precipitation data maps by interpolating the raw point data. And finally in the same way, the soil water storage is a required input in the SWB, described by the Total Available Water (TAW) parameter defined as the amount of water that a crop can extract from its root zone (Allen et al., 1998). Mapping TAW utilizes soil hydraulic properties from the European Soil Database (Panagos et al., 2012), and maximum root depth, which depends in crop type and maximum allowed depth till impervious rock.

3.4 HidroMORE®, a tool for remote sensing based soil water balance (RS-SWB) over large scales

As described previously, the soil water balance turns into a RS-SWB regarding the use of dense NDVI time series, in order to monitor crop biophysical conditions. However, as the RS-SWB is perform at daily time scale, NDVI synthetic images are created, by means of linear temporal interpolation for those dates where no image is available. Furthermore, the dual crop coefficient approach applied, deals with real vegetation conditions after taking into account daily soil water content available for crops. Consequently, a daily monitoring on vegetation evolution development is afforded per each pixel, and thus, achieving local crop biophysical conditions into the soil water balance.

The tool used, able to compute such RS-SWB for the whole Spanish part of Iberian Peninsula, at pixel size scale of $10\,\mathrm{m} \times 10\,\mathrm{m}$, is the HidroMORE® software (Moreno et al., 2017; Torres et al., 2010). HidroMORE® estimates the NIWR that maintains the crop in optimum conditions. Crop optimum conditions means that soil water content is not allowed to be lower than a predetermined threshold, avoiding water stress, or maintaining this water stress on a controlled way, like is the case i.e. for vine grapes. Besides, computation estimates as well the other SWB components like adjusted crop evapotranspiration (ETcadj), run-off (RO), and deep percolation (DP) at same space and time scales. Fig-

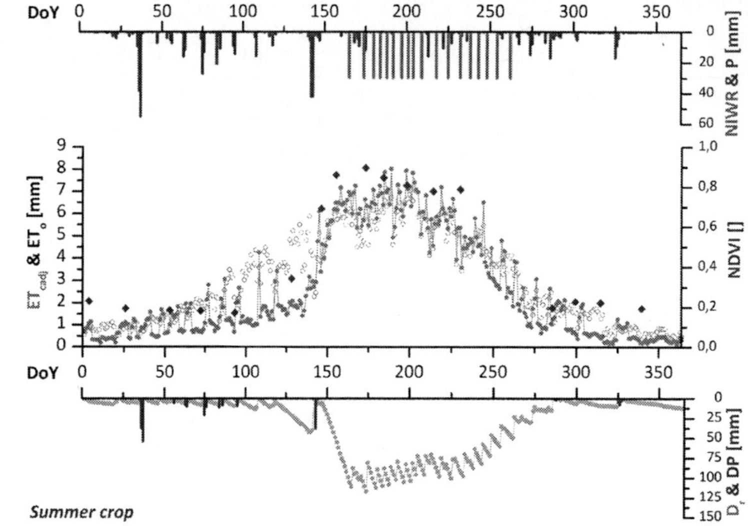

Figure 1. Daily evolution of the different components estimated by the RS-SWB computed by HidroMORE® tool.

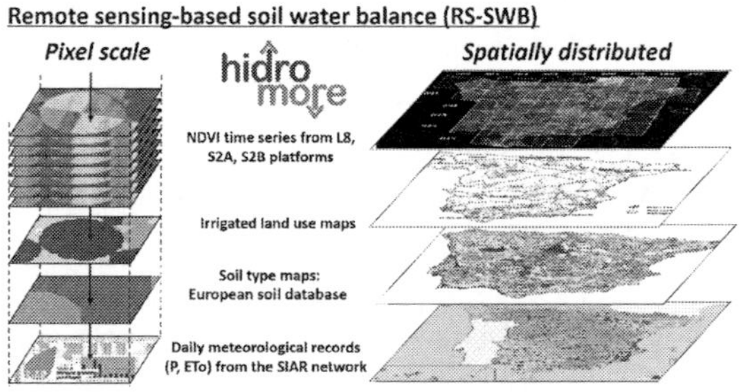

Figure 2. Inputs databases used in HidroMORE® to calculate the RS-SWB at pixel scale and spatially distributed.

ure 1 shows the computation of SWB components, for one pixel.

Inputs for HidroMORE® come from different databases (Fig. 2) like satellite images from L8, S2A and S2B platforms, meteorological daily records from the SIAR network and soil types from the European Soil Database.

3.5 Products accessibility via WebGIS platform

Thematic maps of irrigated areas and its NIWR, are generated at pixel scale, what means a spatial resolution of 0.1 ha, and for NIWR, at daily step. For operational purposes, NIWR data are aggregated at monthly scale. All these products can be online accessed through the SPIDERwebGIS® platform. This platform allows the visualization and display of results, especially as regards the time series. The user must only include the login *demo* and password *demo* through the following link: http://maps.spiderwebgis.org/login/?custom=spider-siar (last access: 13 September 2018).

4 Results of remote sensing-based Irrigation Water Accounting for the mainland Spain

Regarding operational and water management purposes, results from the RS-SWB are aggregated to monthly and annual frequency with the sum of the results estimated at daily time scale. Selected frequency is the reference one used by water managers for governance purpose at the river basin scales. In consequence, 48 monthly thematic maps of NIRW (and other previous cited SWB components), are shown in SPIDERwebGIS® platform.

Figure 3 shows a graph taken from the WebGIS platform that plot the time series of NDVI, Precipitation and NIWR over a single pixel herbaceous irrigated class along the four study years. Therefore, water managers can study the components of the RS-SWB over their territory with temporal and spatially distributed information.

Furthermore, a first evaluation of the remote sensing based irrigation water accounting (RS-IWA) computed in the SPIDER-SIAR project is shown in Fig. 4.

After computing the annual NIWR of the four study years, a zonal statistics calculation was performed per year using the mainland Spain river basin districts perimeters extracted from RBMPs official sources. Among the statistics parameters calculated, annual average NIWR value was then converted into the total volume of NIWR per year, using the irrigated area within each river basin. Finally, the four years average value was compared with the annual gross water de-

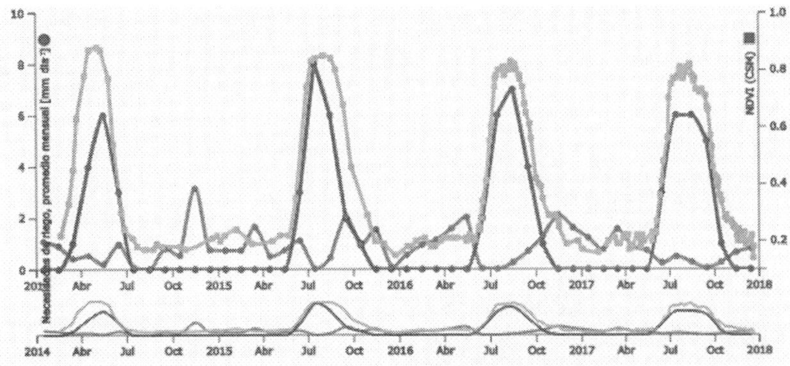

Figure 3. Monthly evolution of the time series of net irrigation water requirements (red chart) and precipitation (green chart), jointly with the NDVI time series (cyan chart) along 2014–2017.

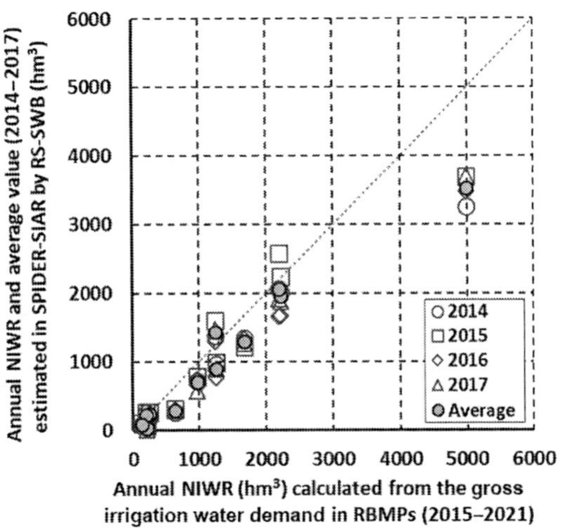

Figure 4. Comparison among the RS-based NIWR and NIWR calculated from official RBMPs for the Spanish mainland along 2014–2017.

mand presented for the river basin water managers in the RBMPs.

Those values were collected from the monitoring and evaluation document about the called by WFD, 2nd cycle River Basin Management Plans (2015–2021), developed by the Spanish Ministry of Agriculture and Fisheries, Food and Environment (MAPAMA and CEDEX, 2017). Then, in order to compare the same irrigation water parameter, the annual gross water demand was converted in NIWR by a global coefficient of 0.65 regarding the water that finally reaches irrigated plots fields, after consider the transport, distribution and application efficiency. The global coefficient assumed, due to the lack of specific data, was selected after consulting the twelve main RBMPs, e.g. Jucar River Basin 0.55, 0.62 in case of Duero River Basin, or 0.76 for the Guadalquivir River Basin.

Figure 4 shows the result for the comparison RS-based NIWR against NIWR calculated from currently official RBMPs. In despite of the encouraging agreement on this comparison, where R^2 value is around 0.95 and RMSE value is 480 hm^3, further analysis will be needed to pave the way to use this technique as an instrument of water planning.

5 Conclusions

The described approach to estimate a remote sensing-based irrigation water accounting (RS-IWA), produces a thematic cartography about irrigated land use and their net irrigation water requirements, spatial and temporally distributed on suitable scales for water governance purposes. The provided information aims to increase water efficiency, help on water governance and management purposes at river basin, regional and national scale, according to the WFD development. The approach requires further analysis about the accuracy and temporal stability of the results.

Data availability. Data sets and results are available through the SPIDERwebGIS® platform. The user must only include the login "demo" and password "demo" through the following link: http://maps.spiderwebgis.org/login/?custom=15spider-siar. For further requirements please contact with corresponding author.

Author contributions. JGR, ACB, LFR, IAA and MBM conceived and conducted the work; JGR and ACB analysed the data; JGR, ACB and ICM wrote and review the article; and RBR and her team at The MAPAMA supervised the project and results provided for the whole period 2014–2017.

Competing interests. The authors declare that they have no conflict of interest.

Special issue statement. This article is part of the special issue "Earth Observation for Integrated Water and Basin Management: New possibilities and challenges for adaptation to a changing environment". It is a result of The Remote Sensing & Hydrology Symposium, Cordoba, Spain, 8–10 May 2018.

Acknowledgements. The authors would like to thank both reviewers for their effort to read and provide comments and suggestions, that led us to improve the content and meaning of the paper.

Edited by: Maria P. González-Dugo
Reviewed by: two anonymous referees

References

Allen, R., Pereira, L. S., Raes, D., and Smith, M.: Crop evapotranspiration – Guidelines for computing crop water requirements, FAO Irrigation and drainage paper 56, Irrigation and Drainage Paper No. 56, FAO, Roma, Italy, 300 pp., 1998.

Anderson, M. C., Neale, C. M. U., Li, F., Norman, J. M., Kustas, W. P., Jayanthi, H., and Chavez, J.: Upscaling ground observations of vegetation water content, canopy height, and leaf area index during SMEX02 using aircraft and Landsat imagery, Remote Sens. Environ., 92, 447–464, https://doi.org/10.1016/j.rse.2004.03.019, 2004.

Balbontín, C., Campos, I., Odi-Lara, M., Ibacache, A., and Calera, A.: Irrigation Performance Assessment in Table Grape Using the Reflectance-Based Crop Coefficient, Remote Sensing, 9, 1276, https://doi.org/10.3390/rs9121276, 2017.

Bausch, W. C. and Neale, C. M. U.: Crop coefficients derived from reflected canopy radiation – A concept, T. ASAE, 30, 703–709, 1987.

Calera, A., Campos, I., Osann, A., D'Urso, G., and Menenti, M.: Remote Sensing for Crop Water Management: From ET Modelling to Services for the End Users, Sensors, 17, 1104, https://doi.org/10.3390/s17051104, 2017.

Campos, I., Neale, C. M. U., Calera, A., Balbontín, C., and González-Piqueras, J.: Assessing satellite-based basal crop coefficients for irrigated grapes (Vitis vinifera L.), Agr. Water Manage., 98, 45–54, https://doi.org/10.1016/j.agwat.2010.07.011, 2010.

Campos, I., Odi, M., Belmonte, M., Martínez-Beltrán, C., and Calera, A.: Obtención de series multitemporales y multisensor de índices de vegetación mediante un procedimiento de normalización absoluta, XIV Congreso de la Asociación Española de Teledetección, 4, 2011.

Campos, I., Villodre, J., Carrara, A., and Calera, A.: Remote sensing-based soil water balance to estimate Mediterranean holm oak savanna (dehesa) evapotranspiration under water stress conditions, J. Hydrol., 494, 1–9, https://doi.org/10.1016/j.jhydrol.2013.04.033, 2013.

Campos, I., Balbontín, C., González-Piqueras, J., González-Dugo, M. P., Neale, C. M. U., and Calera, A.: Combining a water balance model with evapotranspiration measurements to estimate total available soil water in irrigated and rainfed vineyards, Agr. Water Manage., 165, 141–152, https://doi.org/10.1016/j.agwat.2015.11.018, 2016.

Campos, I., Neale, C. M. U., Suyker, A. E., Arkebauer, T. J., and Gonçalves, I. Z.: Reflectance-based crop coefficients REDUX: For operational evapotranspiration estimates in the age of high producing hybrid varieties, Agr. Water Manage., 187, 140–153, https://doi.org/10.1016/j.agwat.2017.03.022, 2017.

Gonzalez-Piqueras, J., Calera, A., Gilabert, M. A., Cuesta, A., and Tercero, F. D. l. C.: Estimation of crop coefficients by means of optimized vegetation indices for corn, Proc. SPIE 5232, Remote Sensing for Agriculture, Ecosystems, and Hydrology V, https://doi.org/10.1117/12.511317, 24 February 2004.

Hornbuckle, J.: Delivering Water Savings through Emerging Technology. In Final Report to Grape and Wine Research & Development Corporation; CSIRO Land and Water: Canberra, Australia, available at: https://www.wineaustralia.com/getmedia/6e7c3e92-c36c-4685-9bf8-9039d99aa4a0/CSL-0901-Final-Report-IrriSAT (last access: 13 September 2018), 2014.

López-Urrea, R., Martínez-Molina, L., de la Cruz, F., Montoro, A., González-Piqueras, J., Odi-Lara, M., and Sánchez, J. M.: Evapotranspiration and crop coefficients of irrigated biomass sorghum for energy production, Irrigation Sci., 34, 287–296, https://doi.org/10.1007/s00271-016-0503-y, 2016.

MAPAMA, and CEDEX: Summary of Spanish River Basin Management Plans of the second cycle, MAPAMA, available at: https://www.mapama.gob.es/es/agua/temas/planificacion-hidrologica/sintesispphh2cicloborrador_tcm30-379039.pdf (last access: 13 September 2018), 2017.

Melton, F. S., Johnson, L. F., Lund, C. P., Pierce, L. L., Michaelis, A. R., Hiatt, S. H., Guzman, A., Adhikari, D., Purdy, A. J., Rosevelt, C., Votava, P., Trout, T. J., Temesgen, B., Frame, K., Sheffner, E. J., and Nemani, R. R.: Satellite Irrigation Management Support With the Terrestrial Observation and Prediction System: A Framework for Integration of Satellite and Surface Observations to Support Improvements in Agricultural Water Resource Management, IEEE J. Sel. Top. Appl., 5, 1709–1721, https://doi.org/10.1109/jstars.2012.2214474, 2012.

Moreno, R., Arias, E., Sánchez, J. L., Cazorla, D., Garrido-Rubio, J., and González-Piqueras, J.: HidroMORE 2: An optimized and parallel version of HidroMORE, 8th International Conference on Information and Communication Systems (ICICS), Irbid, 2017, 1–6, https://doi.org/10.1109/IACS.2017.7921936, 2017.

Neale, C. M. U., Bausch, W. C., and Heerman, D. F.: Development of reflectance-based crop coefficients for corn, T. ASAE, 32, 1891–1899, 1989.

Panagos, P., Van Liedekerke, M., Jones, A., and Montanarella, L.: European Soil Data Centre: Response to European policy support and public data requirements, Land Use Policy, 29, 329–338, https://doi.org/10.1016/j.landusepol.2011.07.003, 2012.

Rouse, J. W., Haas, R. H., Deering, D. W., and Schell, J. A.: Monitoring the vernal advancement and retrogradation of natural vegetation. NASA/GSFC, Greenbelt, USA, Final Report, available at: https://ntrs.nasa.gov/archive/nasa/casi.ntrs.

nasa.gov/19730017588.pdf (last access: 13 September 2018), 1973.

Torres, E. A., Calera, A., González-Piqueras, J., Rubio, E., Campos, I., and Balbotín, C.: Coupling remote sensing and FAO-56 for a distributed water budget in large areas: HidroMORE, Remote Sensing and Hydrology 2010 Symposium, 401–405, 2010.

Proc. IAHS, 380, 37–43, 2018
https://doi.org/10.5194/piahs-380-37-2018

Effect of the water stress on gross primary production modeling of a Mediterranean oak savanna ecosystem

Pedro J. Gómez-Giráldez[1], **Elisabet Carpintero**[2], **Mario Ramos**[2], **Cristina Aguilar**[1], **and María P. González-Dugo**[2]

[1]Fluvial dynamics and hydrology research group, Andalusian Institute of Earth System Research, University of Córdoba, 14071 Córdoba, Spain
[2]IFAPA Consejería de Agricultura Pesca y Desarrollo Rural, Apdo 3048, 14071 Córdoba, Spain

Correspondence: Pedro Jesús Gómez-Giráldez (pjgomezgiraldez@gmail.com)

Received: 17 April 2018 – Revised: 16 July 2018 – Accepted: 19 July 2018 – Published: 18 December 2018

Abstract. *Dehesa* ecosystem consists of widely-spaced oak trees combined with crops, pasture and Mediterranean shrubs. It is located in the southwest of the Iberian Peninsula, where water scarcity is recurrent, severely affecting the multiple productions and services of the ecosystem. Upscaling in situ Gross Primary Production (GPP) estimates in these areas is challenging for regional and global studies, given the significant spatial variability of plant functional types and the vegetation stresses usually present. The estimation of GPP is often addressed using light use efficiency models (LUE-models). Under soil water deficit conditions, biomass production is reduced below its potential rate. This work investigates the effect of different parameterizations to account for water stress on GPP estimates and their agreement with observations. Ground measurements of GPP are obtained using an Eddy Covariance (EC) system installed over an experimental site located in Córdoba, Spain. GPP is estimated with a LUE-model in the footprint of the EC tower using several approaches: a fixed value taken from previous literature; a fixed value modified by daily weather conditions; and both formulations modified by an additional coefficient to explicitly consider the vegetation water stress. The preliminary results obtained during two hydrological years (2015/2016 and 2016/2017) are compared, focusing on specific wet and dry periods.

1 Introduction

Dehesa (known as *montado* in Portugal) ecosystem combines forest, agricultural and extensive livestock productions, presenting important ecosystem services and cultural values. It is composed of sparse trees (mainly holm oak) and an undergrowth of shrub, pasture or herbaceous crop, constituting a characteristic landscape of the southwest of the Iberian Peninsula (Parsons, 1962).

This landscape is characterized by the low fertility of the soils, not suitable for regular farming, and a Mediterranean climate with a high vulnerability to global warming, with increasingly extreme droughts and torrential rainfalls (Kovats et al., 2014).

The assimilation of CO_2 due to the vegetation is represented by the gross primary production (GPP). This production is often estimated from remote sensing based on the works of Monteith (1972) that use biophysical variables and subsequently validated with eddy covariance (EC) systems (e.g. Migliavacca et al., 2009; Wagle et al., 2014; Zhang et al., 2015). These models, known as light use efficiency (LUE) models, relate the incident solar radiation with the photosynthetic activity of the plant, or canopy, through a LUE parameter, which is the amount of biomass produced per unit of radiation absorbed. Under soil water deficit conditions, biomass production is reduced below its potential rate, but this effect is sometimes addressed only indirectly by these models.

The objective of this work is to test different approaches to consider water stress on GPP estimates over a *dehesa* ecosystem using a LUE-model. GPP has been estimated with a LUE-model using field data, Sentinel-2 images, meteorological information and several LUE approaches: a fixed value;

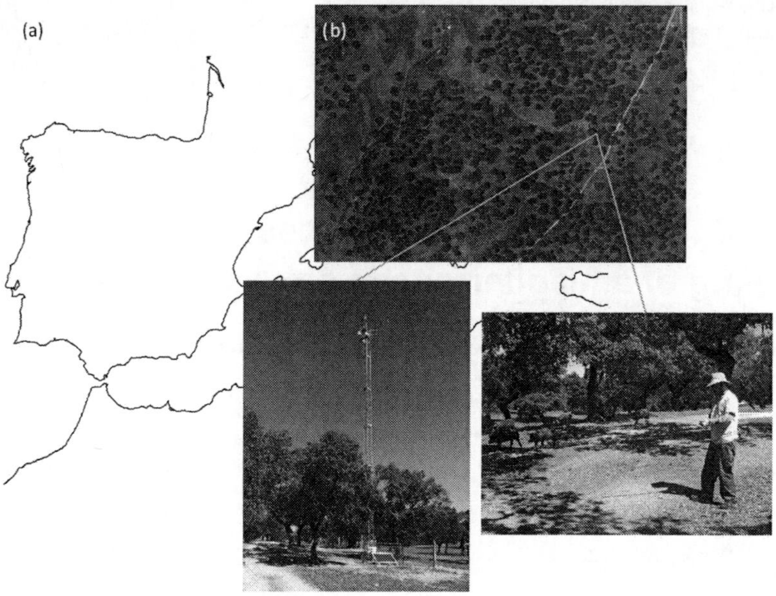

Figure 1. Study site **(a)**, Aerial photography of the area **(b)**, EC tower **(c)** and field measurements example **(d)**.

a fixed value modified by daily weather conditions; and both formulations with an additional coefficient to explicitly consider the vegetation water stress. The results were compared to those obtained from an EC system from July 2015 to September 2017, analyzing the behavior of the selected approaches at different temporal scales.

2 Material and methods

2.1 Study site and eddy covariance data

The study site is a farm located in Cardeña (Córdoba, Spain) (Fig. 1a), part of an environmental protected area. It presents an average rainfall of 720 (±150) mm per year, cold winters, long and dry summers and periodic severe droughts.

The EC system is installed in a gently sloped area with uniform fetch (length in the prevailing wind direction) and homogeneous vegetation (holm oak and pasture) (Fig. 1b and c). The equipment is located at 18 m above ground level in order to minimize the effect of roughness (trees present 7-8 meters of height). A limited number of available measurements caused by a loss of data of the EC tower during one month occurred in the first trimester of 2016. GPP was estimated using Eq. (1):

$$\mathrm{GPP} = \mathrm{NEE} - R_{\mathrm{eco}} \tag{1}$$

where NEE is the net ecosystem exchange given by the EC and R_{eco} is the respiration of the heterotrophic part of the ecosystem which was calculated by a day-time based flux-partitioning algorithm (Lasslop et al., 2010).

2.2 Application of a LUE-model

The GPP was estimated using an adaptation of Monteith (1972) model (Eq. 2):

$$\mathrm{GPP} = \mathrm{fPAR}\ \mathrm{PAR}\varepsilon \tag{2}$$

where GPP ($\mathrm{g\,m^{-2}}$) is the gross primary production, fPAR (dimensionless) is the fraction of photosynthetically active radiation absorbed by the vegetation, PAR ($\mathrm{MJ\,m^{-2}}$) is the photosynthetically active radiation and ε ($\mathrm{g\,MJ^{-1}}$) is the light use efficiency.

2.2.1 Estimation of fPAR

Sentinel-2 images and field measurements have been combined to estimate fPAR. A set of 55 Sentinel-2 cloud-free images were selected (Fig. 2a) and the NDVI was calculated using bands 4 (red) and 8 (NIR) (Fig. 2b) for the study period. The resulting images were linearly interpolated pixel by pixel to obtain a daily NDVI image with 10 m of spatial resolution.

For the transformation NDVI to fPAR, holm oaks and the pasture were addressed separately. For the holm oak, fPAR was monthly monitored by measuring 15 selected trees from the footprint area of the EC during the study period with a LP-80 ceptometer (Fig. 1d) and computing an average from these values that was used as constant value per month. For the pasture, it was used a linear relationship of NDVI-fPAR determined from satellite data in a previous study over the same area (Gómez-Giráldez et al., 2018). The separation

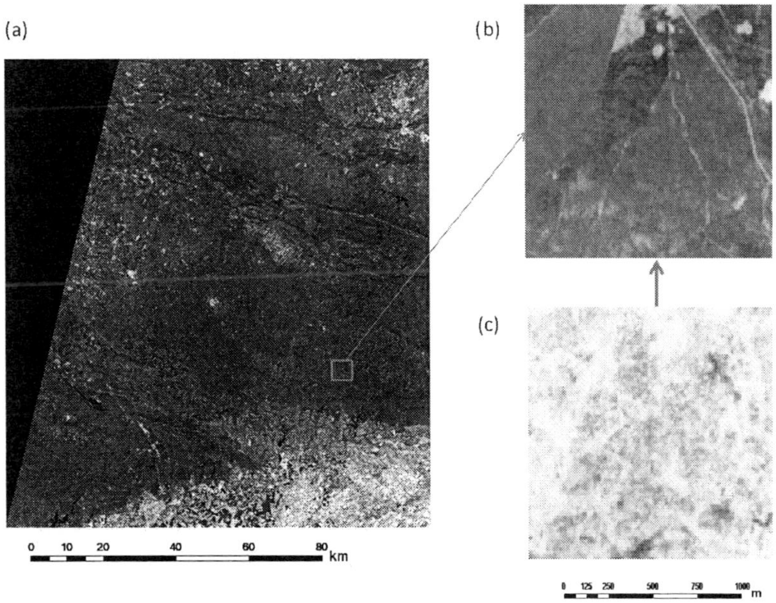

Figure 2. RGB Sentinel-2a image (7 April 2016) **(a)**, NDVI (7 April 2016) **(b)** and fcover images (28 July 2017) **(c)**.

between tree and pasture at pixel level was done assuming that the spectral response of both canopies is additive (e.g. Lu et al., 2003; Blanco et al., 2016). To differentiate the percentage of trees and pasture in each pixel, the coverage fraction image (fcover) was obtained using SNAP toolbox by ESA (http://step.esa.int/main/toolboxes/snap/, last access: 25 July 2018). The Sentinel-2 image of 28 July 2017 was used as reference (Fig. 2c). On this date, the pasture is totally dry and the fcover corresponded only to the tree canopy and can be considered constant.

With these premises, Eq. (3) was obtained:

$$\mathrm{fPAR} = \mathrm{fPAR_{oak}fcover} + \mathrm{fPAR_{pasture}}(1 - \mathrm{fcover}). \quad (3)$$

Finally, to obtain a daily value representative of EC footprint area, a daily footprint function was calculated and used to weigh image pixels. The final data is the sum of the product of fPAR and footprint weights.

2.2.2 Estimation of PAR

The PAR estimation is computed using daily values of solar radiation measured with a 4-component net radiometer NR-1 and a reducing factor of 0.48 according to Szeicz (1974), obtained from measurements in a set of points distributed throughout the globe.

2.2.3 Estimation of ε

A maximum value ($\varepsilon_{\max}$) of 0.77 was selected, based on the results obtained by Running et al. (2000) for wooded grassland. Four methods were analysed:

1. *Max*: the use of $\varepsilon_{\max}$ to test the need to attenuate it.
2. *Meteo*: it uses the attenuation proposed by Running et al. (2000) considering the climatic variables that reduce the efficiency of the plant: minimum daily temperature ($T_{\min}$) and the vapour pressure deficit (VPD). A scalar minimum temperature and the scalar VPD, which are simple linear ramp functions between 0 and 1 derived from the daily values of $T_{\min}$ and VPD. These linear function are obtained using threshold values, where the minimum and maximum value for $T_{\min}$ correspond to 0 and 1 vales of scalar $T_{\min}$ (increasing function) respectively; and minimum and maximum value for VPD correspond to 1 and 0 vales of scalar VPD (decreasing function respectively). The threshold values used were: −8 and 11.39 °C for $T_{\min}$, and 0.65 and 3.1 kPa for VPD.
3. *W*: the use of an attenuation factor due to the water stress of the ecosystem computed by Eq. (4):

$$W = \frac{\mathrm{ET}}{\mathrm{ET_r}} \quad (4)$$

where W (dimensionless) is the water stress coefficient multiplying $\varepsilon_{\max}$ value; ET ($\mathrm{mm\,day^{-1}}$) is the

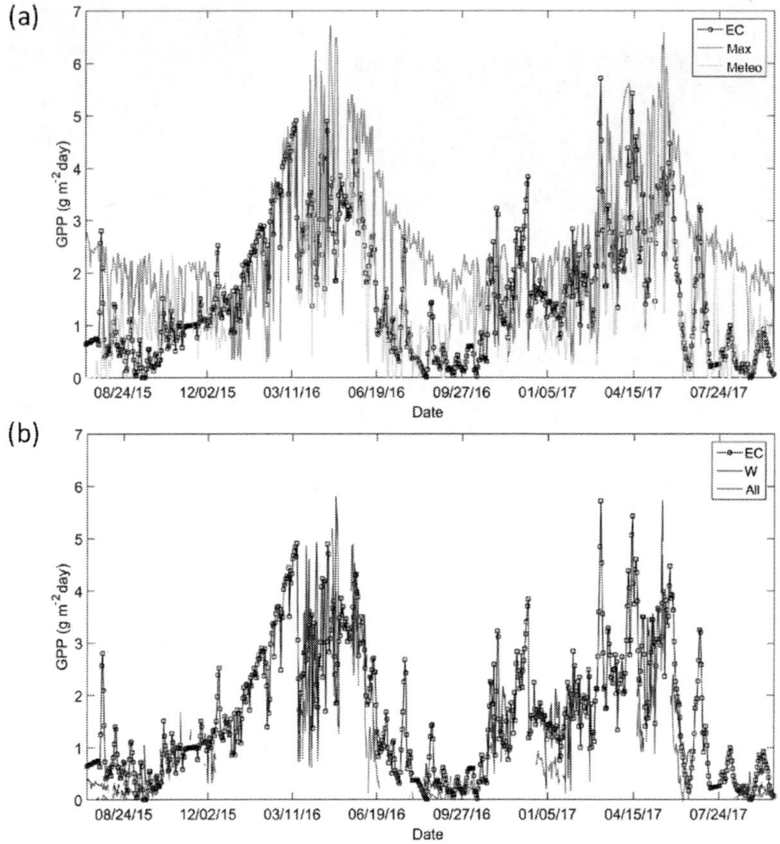

Figure 3. Temporal profile of the GPP fluxes: **(a)** EC, *max* and *meteo* **(b)** EC, *W* and *all*.

daily evapotranspiration of the system by the method of Bowen (1926); and ET_r (mm day^{-1}) is the reference evapotranspiration estimated by the Hargreaves formula (Hargreaves and Samani, 1985).

4. *All*: this approach consider the application of the two previous attenuations, reducing the ε_{max} by multiplying both, W and the scalar VPD and T_{min}.

All the meteorological data required for the calculations were obtained from half-hour measurements of the EC system.

2.3 Study of the methods

In order to study the 4 methods at different temporal scales, the correlation coefficient (R^2) of the linear regressions, the root mean squared error (RMSE) and the Mean Absolute Percentage error (MAPE) of the four methods were calculated at different temporal scales: (i) daily data was analysed for the entire study period; (ii) average values were computed for each of the two hydrological years (1 October to 30 September) 2015/2016 and 2016/2017; (iii) data was averaged in terms of wet (1 October to 15 May) and dry periods (15 May to 30 September).

Finally, the accumulated values for each hydrological year and period were obtained and compared to the measurements of the EC.

3 Results and discussion

The values obtained for fPAR and W for the area vary between 0.2 and 0.7, and 0.03 and 1.10, respectively with average values for the study period of 0.44 and 0.38. Both variables presented a high variability, representative of the Mediterranean climate and vegetation. The average values for ε along the study period were: 0.3 g MJ^{-1} for *meteo* method, 0.29 g MJ^{-1} for W method and 0.14 for *all* method.

Figure 3 shows the temporal profile of GPP at daily scale obtained by the EC and estimated with the four methods considered. The seasonal variation of GPP along the study period can be appreciated, with clear differences between wet and dry periods.

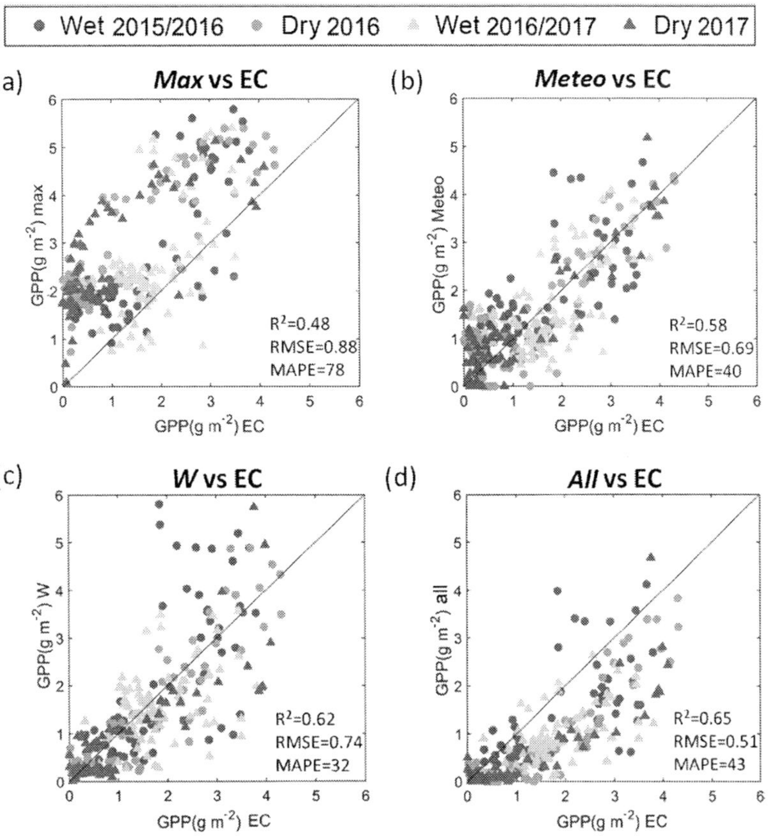

Figure 4. Comparison of GPP measured by the EC vs. max **(a)**, EC vs. meteo **(b)**, EC vs. W **(c)** and EC vs. all **(d)**. Blue: wet period; orange: dry period; circle: 2015/2016; triangle: 2016/2017; black line: 1 : 1 line.

The method *max* (Fig. 4a) presented the weakest correlation (0.48) as expected, the highest error (RMSE = 0.88 g m^{-2}; MAPE = 78 %) and a general overestimation. The *meteo* method (Fig. 4b) improved the results which were similar to those of the W method (Fig. 4c) with no significant biases. Finally, despite the method *all* (Fig. 4d) underestimated the GPP (MAPE = 43 %), it presented the best correlation (0.65) and lower RMSE (0.51 g m^{-2}). The correlation values obtained using W were lower than those obtained previously by Gilabert et al. (2015), but this difference could be explained by the significant higher spatial resolution of this analysis, 10 m, compared to the aggregation at 1 km of their study.

It can be appreciated that there are very low values in summer when the photosynthetic activity is lower. The pasture is dry and not contributing, and the stomata of the trees are often closed. The methods W and *all* performed poorly in this range of very low GPP values. The other cases (*max* and *meteo*) did not present this effect because they overestimated GPP.

The results obtained for both hydrological years (Table 1) were of similar order of magnitude.

The results obtained during the wet periods (Table 1) presented the poorest correlations. The 2016/2017 wet period presented an analogous behaviour, in terms of correlations and errors, compared to that obtained with the whole hydrological year but with lower correlation values. However, the 2015/2016 wet period showed a similar correlation value (around 0.4) in all methods possibly due to the gap in EC.

The results for the dry periods showed the highest correlation (over 0.8 and 0.7 in 2016 and 2017 respectively) but with higher dispersion (higher MAPE than wet period) and the use of the water stress factor improved the results. This difference between the periods confirms the findings of previous studies (e.g. Heinsch et al., 2006).

Finally, Table 2 presents the accumulated GPP values and the differences in percentage with respect to the EC data.

Comparing the accumulated GPP the *max* method strongly overestimated the GPP, in both dry periods (around 150 %) and wet periods (around 50 %). *All* underestimated about a

Table 1. R^2, RMSE ($g\,m^{-2}$) and MAPE (%) obtained by comparison with EC measurements for 2015/2016 and 2016/2017 in the four methods, and over wet and dry periods.

Hydrological year					
2015/2016	R^2	RMSE (MAPE)	2016/2017	R^2	RMSE (MAPE)
max	0.57	0.87 (76)	*max*	0.40	0.86 (82)
meteo	0.58	0.74 (36)	*meteo*	0.59	0.64 (45)
W	0.64	0.84 (34)	*W*	0.62	0.60 (30)
all	0.65	0.59 (42)	*all*	0.68	0.40 (44)
Wet period					
2015/2016	R^2	RMSE (MAPE)	2016/2017	R^2	RMSE (MAPE)
max	0.52	1.15 (43)	*max*	0.28	0.90 (49)
meteo	0.46	0.77 (30)	*meteo*	0.40	0.70 (40)
W	0.47	1.10 (35)	*W*	0.30	0.67 (29)
all	0.44	0.75 (36)	*all*	0.44	0.43 (44)
Dry period					
2016	R^2	RMSE (MAPE)	2017	R^2	RMSE (MAPE)
max	0.77	0.6 (149)	*max*	0.65	0.75 (158)
meteo	0.80	0.41 (47)	*meteo*	0.70	0.61 (48)
W	0.84	0.52 (31)	*W*	0.73	0.53 (39)
all	0.84	0.41 (48)	*all*	0.75	0.40 (55)

Table 2. Accumulated GPP ($g\,m^{-2}\,day^{-1}$) for hydrological year and wet and dry period. The percentage represents the difference respect to EC data.

Hydrological year					
2015/2016	GPP	%	2016/2017	GPP	%
EC	273.9		EC	238.7	
max	562.2	+105.2	*max*	473.9	+98.5
meteo	282.4	+3.1	*meteo*	245.6	+2.9
W	248.4	−9.3	*W*	204.9	−14.2
all	147.0	−46.3	*all*	117.1	−50.9
Wet period					
2015/2016	GPP	%	2016/2017	GPP	%
EC	133.6		EC	160.0	
max	214.2	+60.4	*max*	232.2	+45.1
meteo	133.2	−0.3	*meteo*	142.3	-11.0
W	138.2	−3.4	*W*	137.4	+14.1
all	85.7	−35.8	*all*	82.2	−48.6
Dry period					
2016	GPP	%	2017	GPP	%
EC	101.7		EC	72.5	
max	241.3	+137.3	*max*	198.2	+173.4
meteo	118.6	+16.6	*meteo*	82.7	+14.1
W	96.7	-4.9	*W*	61.8	−14.8
all	57.0	−43.9	*all*	32.1	−55.7

45 % independently dry or wet period. *Meteo* attenuation results obtained for hydrological year and especially for wet period were the most accurate (11 % or less). *W* attenuation was the most accurate for dry periods (differences lower than 15 %), results that agree with other studies as Dong et al. (2015).

4 Conclusions

This study explored different approaches to estimate GPP considering environmental factors and present a preliminary assessment of four different methods. On a daily scale, the combined use of a water stress index and meteorological attenuation provided better GPP estimates regarding R^2 and RMSE than the other formulations, especially for the dry season. However, this method presented a bias at daily scale that for the accumulated values resulted in a significant underestimation of seasonal values.

The use of a fixed value of ε presented the poorest results, supporting the need for environmental attenuation factors, as the approaches tested here. *Meteo* and *W* provided similar estimates, even when the nature of the attenuation considered was different. Further analysis of these factors is required to propose a specific model suited to *dehesa* ecosystem.

Data availability. Data sets are available upon request by contacting the correspondence author.

Author contributions. PJG-G and MPG-D conceived and designed the experiments; PJG-G, EC and MR performed the experiments; PJG-G analysed the data and wrote the article and CA and MPG-D contributed to the interpretation of data and reviewed the article.

Competing interests. The authors declare that they have no conflict of interest.

Special issue statement. This article is part of the special issue "Earth Observation for Integrated Water and Basin Management: New possibilities and challenges for adaptation to a changing environment". It is a result of The Remote Sensing & Hydrology Symposium, Cordoba, Spain, 8–10 May 2018.

Acknowledgements. This work has been funded byLIFE+bioDehesa project (LIFE11/BIO/ES/000726) and INIA-FEDER 2014–2020 (Operational Programme Smart Growth) project RTA2014-00063-C04-02.

Edited by: Rafael Pimentel
Reviewed by: three anonymous referees

References

Blanco, L. J., Paruelo, J. M., Oesterheld, M., and Biurrun, F. N.: Spatial and temporal patterns of herbaceous primary production in semi-arid shrublands: a remote sensing approach, J . Veg. Sci., 27, 716–727, 2016.

Bowen, I. S.: The ratio of heat losses by conduction and by evaporation from any water surface, Phys. Rev., 27, 779–787, 1926.

Dong, J., Xiao, X., Wagle, P., Zhang, G., Zhou, Y., Jin, C., Torn, M. S., Meyers, T. P., Suyker, A. E., Wang, J., Yan, H., Biradar, C., and Moore, B.: Comparison of four EVI-based models for estimating gross primary production of maize and soybean croplands and tallgrass prairie under severe drought, Remote Sens. Environ. 162, 154–168, 2015.

Gilabert, M. A., Moreno, A., Maselli, F., Martínez, B., Chiesi M., Sánchez-Ruiz, S., García-Haro, F. J., Pérez-Hoyos, A., Campos-Taberner, M., Pérez-Priego, O., Serrano-Ortiz, P., and Carrara, A.: Daily GPP estimates in Mediterranean ecosystems by combining remote sensing and meteorological data, ISPRS J. Photogramm., 102, 184–197, 2015.

Gómez-Giráldez, P. J., Aguilar, C., Caño, A. B., García, A., and González-Dugo, M. P.: Remote sensing estimate of net primary production as monitoring indicator of holm oak savanna management, Ecol. Indicat. J., in review, 2018.

Hargreaves, G. H. and Samani, Z. A.: Reference crop evapotranspiration from temperature, Appl. Eng. Agr., 1, 96–99, 1985.

Heinsch, F. A., Zha, M., Running, S. W., Kimball, J. S., Nemani, R. R., Davis, K. J., and Bolstad, P. V.: Evaluation of remote sensing based terrestrial productivity from MODIS using regional tower eddy flux network observations, IEEE T. Geosci. Remote, 44, 1908–1925, 2006.

Kovats, R. S., Valentini, R., Bouwer, L. M., Georgopoulou, E., Jacob, D., Martin, E., Rounsevell, M., and Soussana, J. F.: Climate Change 2014: Impacts, Adaptation, and Vulnerability. Part B: Regional Aspects, in: Contribution of Working Group II to the Fifth Assessment Report of the Intergovernmental Panel on Climate Change, edited by: Barros, V. R., Field, C. B., Dokken, D. J., Mastrandrea, M. D., Mach, K. J., Bilir, T. E., Chatterjee, M., Ebi, K. L., Estrada, Y. O., Genova, R. C., Girma, B., Kissel, E. S., Levy, A. N., MacCracken, S., Mastrandrea, P. R., and White, L. L., Cambridge University Press, Cambridge, UK and New York, NY, USA, 1267–1326, 2014.

Lasslop, G., Reichstein, M., Papale, D., Richardson, A., Arneth, A., Barr, A., Stoy, P., and Wohlfahrt, G.: Separation of net ecosystem exchange into assimilation and respiration using a light response curve approach: critical issues and global evaluation, Global Change Biol., 16, 187–208, 2010.

Lu, H., Raupach, M., McVicar, T., and Barrett, D.: Decomposition of vegetation cover into woody and herbaceous components using AVHRR NDVI time series, Remote Sens. Environ., 86, 1–18, 2003.

Migliavacca, M., Meroni, M., Busetto, L., Colombo, R., Zenone, T., Matteucci, G., Manca, G., and Seufert, G.: Modeling Gross Primary Production of Agro-Forestry Ecosystems by Assimilation of Satellite-Derived Information in a Process-Based Model, Sensors, 9, 922–942, 2009.

Monteith, J. L.: Solar radiation and productivity in tropical ecosystems, J. Appl. Ecol., 9, 747–766, 1972.

Parsons, J. D.: The Acorn-Hog Economy of the Oak Woodlands of Southwestern Spain, Geogr. Rev., 2, 211–235, 1962.

Running, S. W., Thorton, P. E., Nemani, R., and Glassy, J. M.: Global terrestrial gross and net primary productivity from the Earth Observing System, in: Methods in Ecosystem Science, edited by: Sala, O., Jackson, R., and Mooney, H., Springer Verlag, New York, 44–57, 2000.

Szeicz, G.: Solar radiation for plant growth, J. Appl. Ecol., 11, 617–637, 1974.

Wagle, P., Xiao, X., Torn, M. S., Cook, D. R., Matamala, R., Fischer, M. L., Jin, C., Dong, J., and Biradar, C.: Sensitivity of vegetation indices and gross primary production of tallgrass prairie to severe drought, Remote Sens. Environ., 152, 1–14, 2014.

Zhang, L. X., Zhou, D. C., Fan, J. W., and Hu, Z. M.: Comparison of Four Light Use Efficiency Models for Estimating Terrestrial Gross Primary Production, Ecol. Model., 300, 30–39, 2015.

Proc. IAHS, 380, 45–54, 2018
https://doi.org/10.5194/piahs-380-45-2018

Remote sensing vegetation index methods to evaluate changes in greenness and evapotranspiration in riparian vegetation in response to the Minute 319 environmental pulse flow to Mexico

Pamela L. Nagler[1], **Christopher J. Jarchow**[1], **and Edward P. Glenn**[2,†]

[1]U.S. Geological Survey, Southwest Biological Science Center, 520 N. Park Avenue, Tucson, AZ, USA
[2]Environmental Research Laboratory, University of Arizona, 2601 E. Airport Drive, Tucson, AZ, USA
[†]deceased

Correspondence: Pamela L. Nagler (pnagler@usgs.gov)

Received: 28 April 2018 – Revised: 9 August 2018 – Accepted: 4 September 2018 – Published: 18 December 2018

Abstract. During the spring of 2014, 130 million m^3 of water were released from the United States' Morelos Dam on the lower Colorado River to Mexico, allowing water to reach the Gulf of California for the first time in 13 years. Our study assessed the effects of water transfer or ecological environmental flows from one nation to another, using remote sensing. Spatial applications for water resource evaluation are important for binational, integrated water resources management and planning for the Colorado River, which includes seven basin states in the US plus two states in Mexico. Our study examined the effects of the historic binational experiment (the Minute 319 agreement) on vegetative response along the riparian corridor. We used 250 m Moderate Resolution Imaging Spectroradiometer (MODIS), Enhanced Vegetation Index (EVI) and 30 m Landsat 8 satellite imagery to track evapotranspiration (ET) and the normalized difference vegetation index (NDVI). Our analysis showed an overall increase in NDVI and evapotranspiration (ET) in the year following the 2014 pulse, which reversed a decline in those metrics since the last major flood in 2000. NDVI and ET levels decreased in 2015, but were still significantly higher ($P < 0.001$) than pre-pulse (2013) levels. Preliminary findings show that the decline in 2015 persisted into 2016 and 2017. We continue to analyse results for 2018 in comparison to short-term (2013–2018) and long-term (2000–2018) trends. Our results support the conclusion that these environmental flows from the US to Mexico via the Minute 319 "pulse" had a positive, but short-lived (1 year), impact on vegetation growth in the delta.

1 Introduction

In 2012, Minute 319 to the 1944 Treaty was signed (https://www.ibwc.gov/Treaties_Minutes/Minutes.html, last access: 6 September 2018) allowing for a pulse flow of water into the Colorado River's delta, in Mexico. During the spring of 2014 (23 March to 18 May), 130 million m^3 of water were released from the United States' Morelos Dam on the lower Colorado River to Mexico, allowing water to reach the Gulf of California for the first time in 13 years. Our study assessed the effects of water transfer via ecological environmental flows from one nation to another, using remote sensing of vegetative response. Spatial applications for water resource evaluation are important for binational, integrated water resources management and planning for the Colorado River, which includes seven basin states in the US plus two states in Mexico.

1.1 Background

Water is a primary environmental driver of plant productivity in our region and an important metric of the efficiency of water use is evapotranspiration (ET). As with other similar regions of the world, ET is a key component of the hydrological cycle and can indicate the resilience and resistance of

Published by Copernicus Publications on behalf of the International Association of Hydrological Sciences.

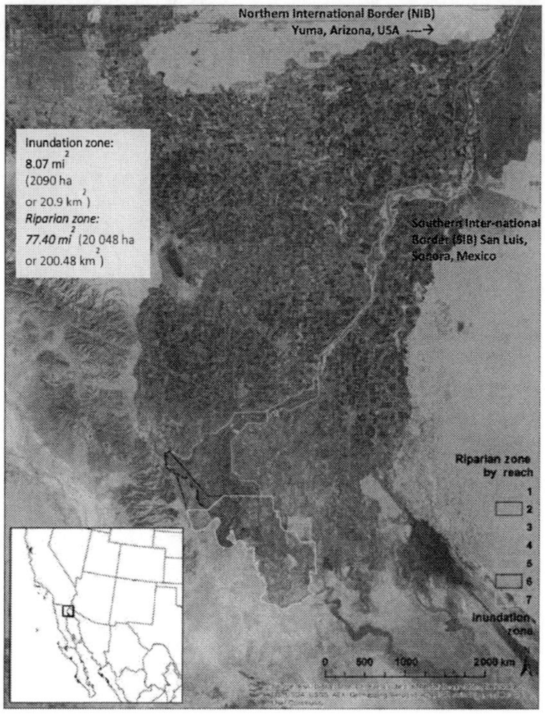

Figure 1. Map of the Colorado River Delta showing the seven reaches of the Riparian Corridor in Mexico.

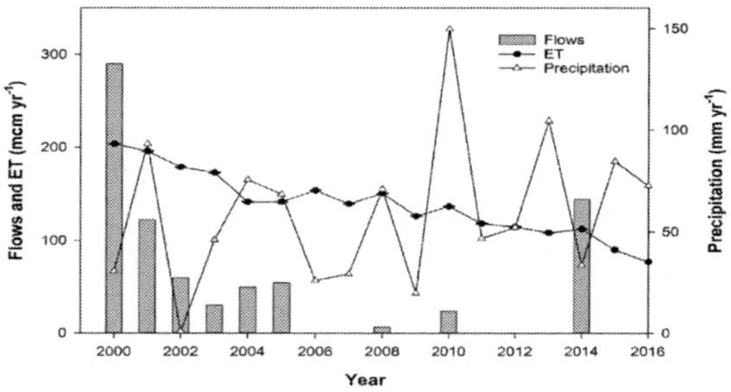

Figure 2. Annual flow volume and precipitation at the Southern International Boundary (SIB) and total annual ET (based on MODIS EVI) within the riparian corridor from 2000–2016.

vegetation to drought. In undisturbed deserts, nearly all the water arriving as precipitation can be discharged as ET, but landcover disturbances can greatly impact the water balance. For example, beetle infestations can lower vegetation cover, scouring flows can clear vegetation, and fire and burn scars can increase water use by cohorts of older trees (Bateman et al., 2013; Nagler et al., 2013c). Understanding how intact plant communities utilize water and how disturbance alters water-use efficiency can help predict future ecosystem resilience (Shanafield et al., 2017). Even in sparse deserts, vegetation is often the most important factor controlling how water is utilized or lost in ecosystems (Gee et al., 1994; Glenn et al., 2008). Measuring plant water use or landscape-level ET is important primarily because water retention and efficient utilization is critical for the survival of humans, croplands, and natural areas. The preservation of natural areas protects water resources and the plants and animals that use those areas as habitat (Hinojosa-Huerta et al., 2013).

1.2 Research goal

Our study examined the effects of the historic binational experiment (the Minute 319 agreement) on vegetative response, greenup, and ET along the riparian corridor. The purpose of these environmental pulse flows was to demonstrate changes in biology and hydrology processes, but we hoped to see it recharge groundwater and germinate new cohorts of riparian plants to restore the riparian plant communities along the delta. Flows were designed to simulate historical (natural) pulse-flows (i.e., prior to dams and river regulations altering the flow regime). Our goals were to: (1) estimate ET using a Moderate Resolution Imaging Spectrometer (MODIS) Enhanced Vegetation Index (EVI)-based ET algorithm and (2) assess the greenup response of vegetation using Landsat 8 Normalized Difference Vegetation Index (NDVI).

2 Methods

We document the changes in ET and green foliage density attributable to the Minute 319 Pulse flow and compare them with conditions from 2000–2013, a drought period that preceded the pulse flow (Jarchow et al., 2017b). Landsat imagery (30 m resolution, 16-day return time) and MODIS imagery (250 m resolution, daily return time) were used for the analyses. The analyses used vegetation indices, which ratio different optical bands to provide a measure of canopy "greenness". We used the NDVI for Landsat images while the EVI was used for MODIS.

We used the entire MODIS EVI time-series, every 16 days since February of 2000 to present; approximately 335 MODIS images were used in this study. We used 4–5 clear sky summer Landsat images per year for the assessment during the years 2013–2018. At least five Landsat scenes were acquired during the growing season (May–October) for the years 2013, 2014, 2015, 2016 and 2017 and NDVI was averaged across the growing season. These imagery selection choices were based on previous performance comparisons made in riparian ecosystems (Groeneveld and Baugh, 2007; Groeneveld et al., 2007; Nagler et al., 2013). MODIS EVI and potential ET from the Yuma Valley AZMET station were used to estimate ET with an algorithm previously calibrated with eddy covariance flux tower data in riparian ecosystems (Nagler el al., 2005a, b) and we used this ET equation which was revised for dryland riparian and agricultural lands that

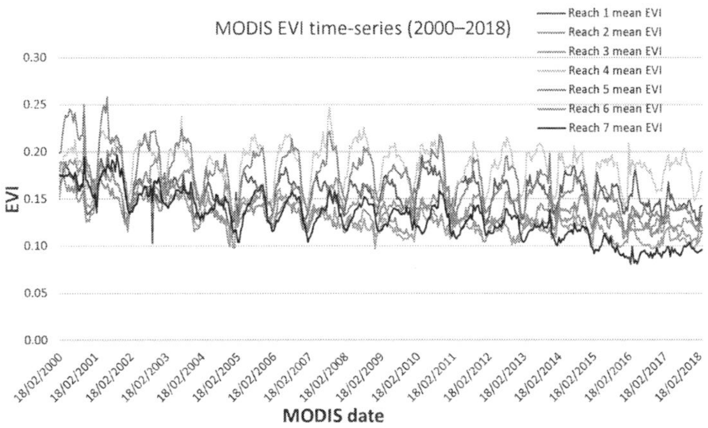

Figure 3. MODIS EVI time series from 2000 to 2018 for the seven Reaches of the riparian corridor.

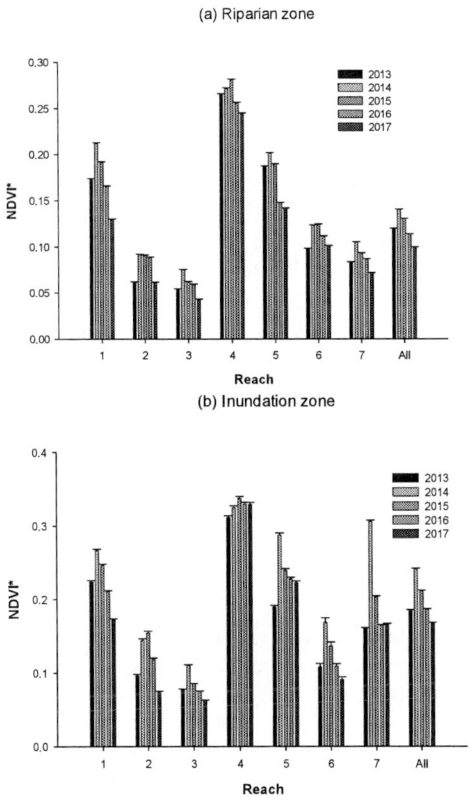

Figure 4. (a) Landsat NDVI in the entire riparian zone, by river Reach, from 2013–2017. **(b)** NDVI in the inundation zone, by river Reach, from 2013–2017. The 2013–2015 NDVI data was derived from Jarchow et al. (2017).

included similar areas to the Colorado River Delta in Mexico, including those in the arid, southwestern US (Nagler et al., 2013). The resulting algorithm capably predicted ETa across riparian plants and crops ($r^2 = 0.73$). Equation (1), which has a biophysical justification based on the Beer-Lambert Law, succeeds in unifying riparian and crop data into a single algorithm:

$$\mathrm{ETa} = \mathrm{ETo}[1.73(1 - e^{-2.25}\mathrm{EVI}) - 0.220] \quad (1)$$

The riparian corridor was divided into seven river Reaches from the Northern International Border (NIB) near Yuma, AZ to the estuary (Fig. 1). Landsat NDVI ($n = 4$ to 5 summer season scenes/year) was compared for each of the seven river Reaches for the wetted, inundation zone and the overall riparian corridor.

3 Results

Monitoring of the riparian corridor was done using both MODIS EVI at the coarser level as well as Landsat NDVI at the finer level, and provided the greenness trends over the long-term (nearly two decades) and conditions of the vegetation cover over the short-term, since the year prior to the Minute 319 pulse flow through March 2018.

Figure 2 shows the annual flow volume and precipitation at the Southern International Boundary (SIB) and total annual ET (based on MODIS EVI within the riparian corridor) from 2000–2016. The last large flow in the year 2000 was approximately twice that of the Minute 319 flow.

A declining trend since the year 2000 for all the Reaches is noted, with the exception of Reach 4 which contains a variety and number of restoration plots, planted by field crews working to restore native vegetation habitat in the delta. Figure 3 shows MODIS EVI from 2000 to present for the riparian corridor for each of the seven Reaches. Figure 4 shows finer res-

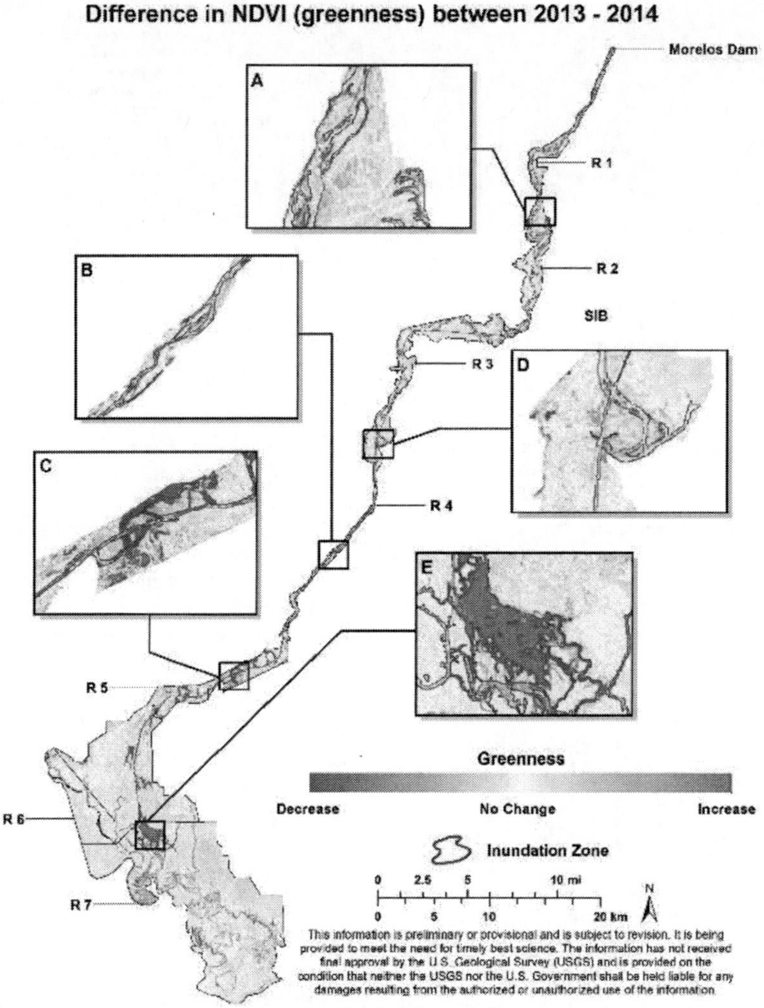

Figure 5. 2014 pulse flow inundation zone and the difference in NDVI from 2013 and 2014. Greener color indicates higher NDVI than in previous year; browner color indicates lower NDVI than in previous year. Image from Jarchow et al. (2017a).

olution Landsat images in which NDVI was acquired by river Reach. Then change maps of NDVI or "greenness" were created. Increases in NDVI in 2014 occurred in the zone inundated by the pulse flow as well as in the non-inundated outer parts of the riparian floodplain, where groundwater supported existing vegetation. NDVI was markedly higher in 2014 than in 2013 for all Reaches except Reach 4, where plants were cleared prior to the pulse flow. The overall NDVI increase above the soil baseline value (0.0947) was 17 % and was statistically significant ($P < 0.001$). NDVI decreased in each year since 2014 with the exception of the restoration areas, for both the riparian zone and the inundation zone (Fig. 4). Figure 4 also shows that from 2016–2017 NDVI decreased steadily, falling below 2013 levels in the riparian corridor. The most intense greening in 2014 took place in the zone of inundation but increases in NDVI also occurred outside the zone of inundation, indicating that the pulse water was stored as groundwater in these outside areas. Our analysis showed an overall increase in NDVI and evapotranspiration (ET) in the year following the 2014 pulse, which reversed a decline in those metrics since the last major flood in 2000. NDVI and ET levels decreased in 2015, but were still significantly higher than pre-pulse (2013) levels. Preliminary findings show that this decline persisted into 2016 and 2017.

Figure 5 shows areas inundated during the pulse flow and differences in NDVI between 2013 (pre-pulse) and 2014

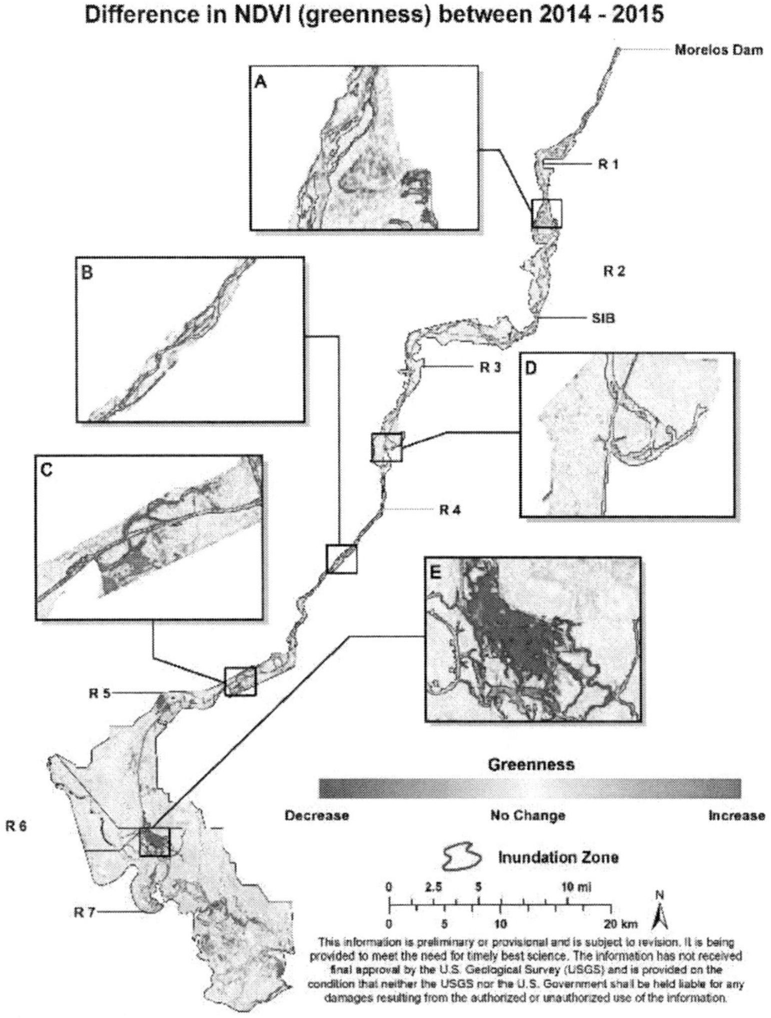

Figure 6. 2014 pulse flow inundation zone and the difference in NDVI from 2014 and 2015. Greener color indicates higher NDV than in previous year; browner color indicates lower NDVI than in previous year.

(post-pulse), with selected enlarged portions of the riparian corridor. A greener color indicates that NDVI was higher in 2014 than in 2013. There was extensive green-up in all areas, except for the portion in the lower part of Reach 4 (Fig. 5c), where extensive land-clearing took place prior to the pulse flow. Much of the land cleared was not inundated during the pulse flow.

Figure 6 shows areas inundated during the pulse flow and differences in NDVI between 2014 and 2015. A greener color indicates that NDVI was higher in 2015 than in 2014. A browner color indicates a reduction in greenness (not necessarily the result of brown vegetation) from 2014 to 2015. Note that while some areas were greener than in the post-pulse growing season of 2014 (Fig. 6a and c), other parts of the riparian corridor were not as green as in the previous year – see especially enlarged part of Reach 7 (Fig. 6e).

Figure 7 shows areas inundated during the pulse flow and differences in NDVI between 2013 (pre-pulse) and 2015 (two growing seasons after the pulse flow). Some areas continued to increase in greenness from 2013 to 2015 (lower Reach 1 and Reach 7), while other areas show little change, or were less green than under pre-pulse conditions.

Figure 8 shows areas inundated during the pulse flow and differences in NDVI between 2015 and 2016. Note that the overall trend was a decrease in greenness, but some local-

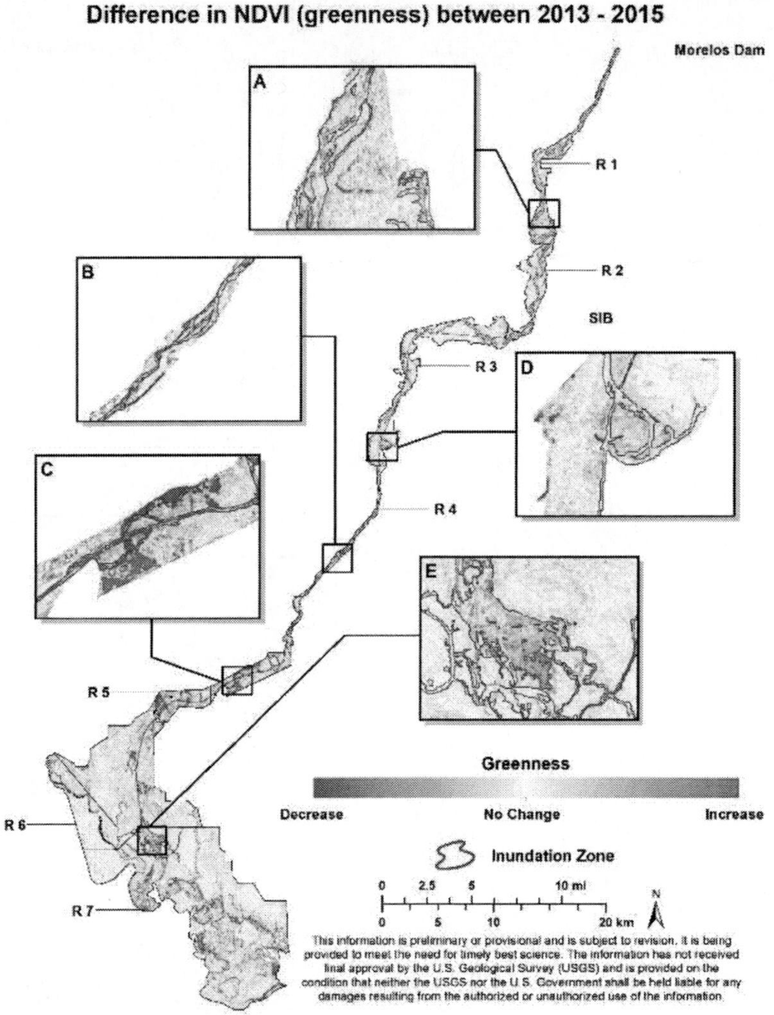

Figure 7. 2014 pulse flow inundation zone and the difference in NDVI from 2013 and 2015. Greener color indicates higher NDVI than in previous year; browner color indicates lower NDVI than in previous year.

ized areas (such as in Reach 7 and Reach 3; Fig. 8e and d, respectively) displayed a slight increase in greenness.

Figure 9 shows areas inundated during the pulse flow and differences in NDVI between 2016 and 2017. Note that the overall trend was a decrease in greenness in 2017, but the area corresponding to the inundation zone in Reach 7 (Fig. 9e) saw a slight increase in greenness.

4 Discussion

We have analyzed the effects of this historic release of water, providing insight into the efficacy of environmental pulses as a tool for restoring the delta's riparian corridor, an ecosystem heavily affected by decades of impoundments and diversions.

Landsat NDVI was averaged across the growing season (May–October) from 2013–2017 for each river Reach and all Reaches combined. NDVI is greatest in Reaches 1, 4 and 5, where shallow groundwater and surface water supports vegetation. Reaches 2 and 3 are within the "dry Reach" where the water table is deep and vegetation is sparse. Reach 6 is dominated by the Río Hardy drainage and was largely unaffected by the pulse flow and subsequent base flow. Reach 7 includes the upper estuary and received surface water from the pulse flow in 2014 and more regular flows from the. Groundwater is shallow in Reach 7.

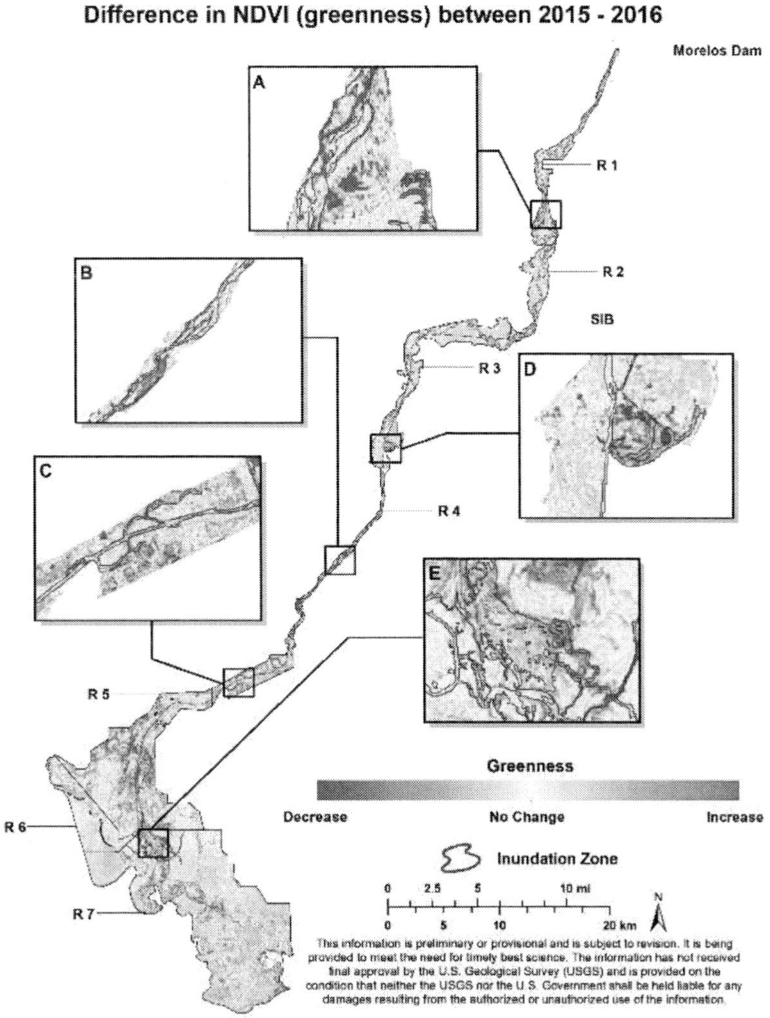

Figure 8. 2014 pulse flow inundation zone and the difference in NDVI from 2015 and 2016. Greener color indicates higher NDVI than in previous year; browner color indicates lower NDVI than in previous year.

NDVI was higher in 2014 than in 2013 for all Reaches. The overall NDVI increase from 2013 to 2014 was 17 % ($P < 0.001$). The most intense greening in 2014 took place in the zone of inundation by the pulse flow, but increases in NDVI also occurred outside the zone of inundation, indicating that the pulse flow likely enhanced groundwater conditions in those areas as well.

The overall peak NDVI values occurred in Reach 4 in 2015, perhaps reflecting the effects of planting and vegetation growth in the Laguna Grande restoration site.

For Reaches 1, 4, 5, and all combined, NDVI decreased steadily from 2016–2017, falling below 2013 levels. The rapid decrease in NDVI values in Reach 1, and the 2017 drop in the Reach 2 may be consequences of the expanding cone of depression on the water table noted in Section 2 of this report. By 2017, NDVI values in the Reaches 2 and 3 – the dry Reaches- and Reaches 6 (Río Hardy) and 7 (the upper estuary) fell to values similar to or slightly lower than those observed in 2013. Restoration activities at the Miguel Aleman site appear to have been at too small a scale to sustain overall NDVI values in Reach 2.

The Reach 4 average NDVI values did not fall as much after 2014 as in other Reaches, perhaps as a consequence of base flow deliveries to the two restoration sites in this Reach, the persistent high water table in this Reach, or, most likely, both factors.

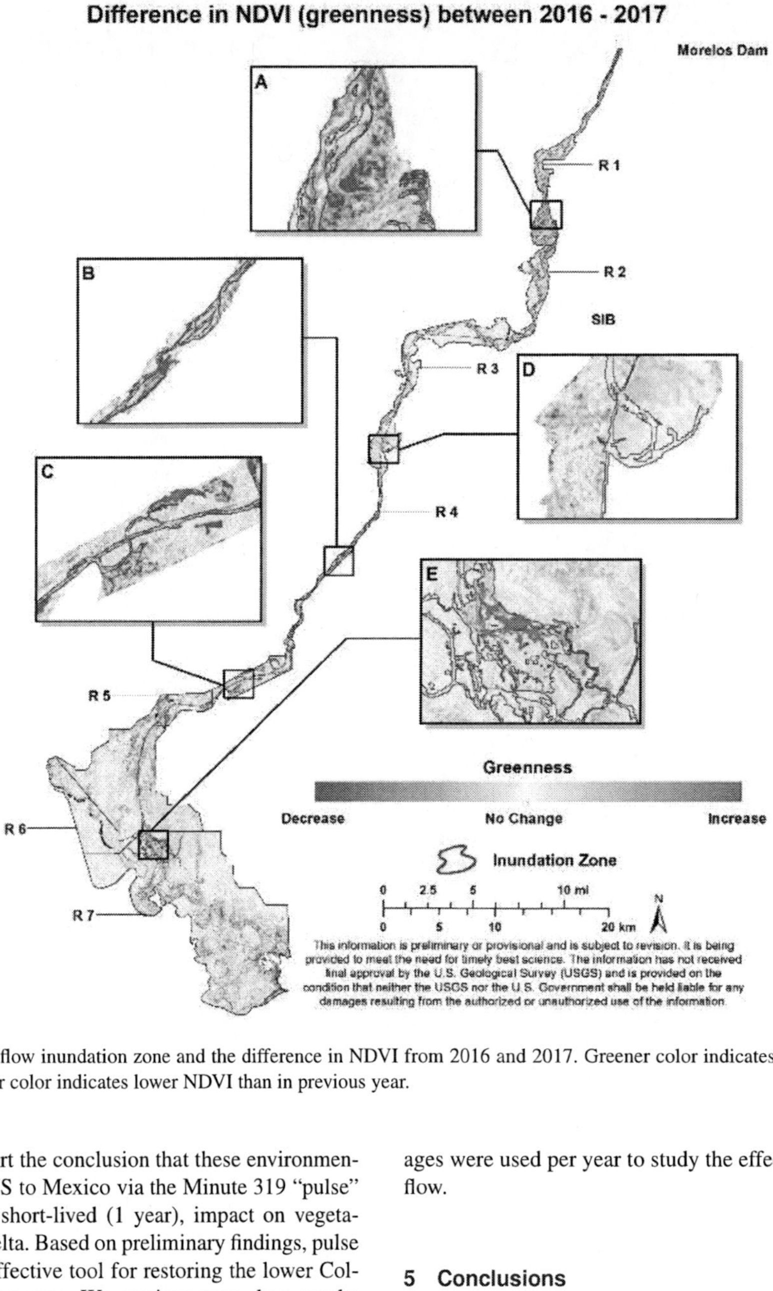

Figure 9. 2014 pulse flow inundation zone and the difference in NDVI from 2016 and 2017. Greener color indicates higher NDVI than in previous year; browner color indicates lower NDVI than in previous year.

Our results support the conclusion that these environmental flows from the US to Mexico via the Minute 319 "pulse" had a positive, but short-lived (1 year), impact on vegetation growth in the delta. Based on preliminary findings, pulse flows could be an effective tool for restoring the lower Colorado River's riparian zone. We continue to analyze results for 2018 in comparison to short-term (2013–2018) and long-term (2000–2018) trends using both MODIS EVI and Landsat NDVI. The long-term trends are able to be assessed using MODIS EVI time-series data, by Reach, for the riparian corridor. We are comparing this longer period to the area we focused on between 2013 (prior to the pulse flow) and the 2014–2018 period in which 4–5 summer Landsat NDVI images were used per year to study the effect of the 2014 pulse flow.

5 Conclusions

The Minute 319 Pulse Flow produced a 17 % increase in NDVI throughout all seven Reaches of the riparian corridor in 2014, compared with 2013. The most intense greening in 2014 took place in the zone of inundation, but increases in NDVI also occurred outside the zone of inundation, indicating that the pulse flow likely enhanced groundwater conditions in those areas as well. For Reaches 1, 4, 5, and all combined, NDVI and greenup steadily declined from 2015–2017, eventually falling to or below 2013 (pre-pulse) levels.

The pulse flow and subsequent base flows did not – at the scale of Reaches, and at 30 m satellite image resolution – produce effects on vegetation greenness in the riparian zone that persisted to the end of the 2017 growing season. Increases in greenness within restoration sites supplied with base flows are not sufficient to maintain the high average, Reach-level, NDVI values observed in the growing season after the 2014 pulse flow. The restoration sites may be too small to have a strong effect on Reach-level averages.

Although NDVI was highest in 2014 after the pulse flow, the magnitude is not what was achieved in previous flood years and it did not persist after 2014. A series of releases and active restoration projects will contribute to helping to rebuild the vegetation community to its former status. Since floods in these Reaches are rare, the bulk of riparian ET appears to be supported by existing underflows of water from the US (Glenn et al., 2001; Nagler et al., 2008). A pulse flow can be expected to have benefits in riparian corridor in the release year, after which it will flow into the subterranean estuary, where it may also perform ecosystem services. Active restoration projects can help make maximum use of environmental flows. These can include existing activities in Reach 4 to restore cottonwood habitat, but could be expanded to include mesquites (*Prosopis* spp.) in Reaches 2 and 3, where depth to groundwater is too deep for cottonwoods and willows, but the groundwater is low salinity and could support mesquites.

Data availability. Data either are not available, or have limited availability, due to proprietary restrictions. Contact Pam Nagler (pnagler@usgs.gov) for more information.

Author contributions. PLN, EPG, CJJ conceived and designed the research; CJJ, PLN performed the experiments and analyzed the data; CJJ, PLN wrote and edited the manuscript.

Competing interests. The authors declare that they have no conflict of interest.

Special issue statement. This article is part of the special issue "Earth Observation for Integrated Water and Basin Management: New possibilities and challenges for adaptation to a changing environment". It is a result of The Remote Sensing & Hydrology Symposium, Cordoba, Spain, 8–10 May 2018.

Acknowledgements. We would like to posthumously honor Edward P. Glenn for his dedication to the Colorado River's delta, a tri-national region including the US, Mexico and Native American Tribal lands. We thank Edward P. Glenn for mentoring dozens of students and helping NGO teams over four decades. Thanks to Edward P. Glenn's passionate interest, his personal and professional investment, and his research contributions as a prolific writer, several environmental Minutes regarding water quality & salinity (306), base flows (319), and restoration activities (323) have been added to the 1944 international treaty (https://www.ibwc.gov/Treaties_Minutes/Minutes.html, last access: 6 September 2018). We wish to thank Karl Flessa from the University of Arizona for his review and text additions. We also wish to thank the Minute 319 Delta Science Team for putting in countless hours of hard work and providing invaluable help on this project. Without their help, this project would not have been possible. Any use of trade, firm, or product names is for descriptive purposes only and does not imply endorsement by the U.S. Government.

Edited by: Rafael Pimentel
Reviewed by: two anonymous referees

References

Bateman, H. L., Nagler, P. L., and Glenn, E. P.: Plot and landscape-level changes in climate and vegetation following defoliation of exotic saltcedar (*Tamarix* sp.) from the biocontrol agent *Diorhabda carinulata* along a stream in the Mojave Desert, USA J. Arid Environ., 89, 16–20, 2013.

Gee, G. W., Wierenga, P. J., Andraski, B. J., Young, M. H., Fayer, M. J., and Rockhold, M. L.: Variations in water balance and recharge potential at three western desert sites, Soil Sci. Soc. Am. J., 58, 63–72, 1994.

Glenn, E. P., Zamora-Arroyo, F., Nagler, P. L., Briggs, M., Shaw, W., and Flessa, K.: Ecology and conservation biology of the Colorado River Delta, Mexico, J. Arid Environ., 49, 5–16, 2001.

Glenn, E. P., Morino, K., Didan, K., Jordan, F., Carroll, K. C., Nagler, P. L., Hultine, K., Sheader, L., and Waugh J.: Scaling sap flux measurements of grazed and ungrazed shrub communities with fine and coarse-resolution remote sensing, Ecohydrology, 1, 316–329, 2008.

Groeneveld, D. P. and Baugh, W. M.: Correcting satellite data to detect vegetation signal for eco-hydrologic analyses, J. Hydrol., 344, 135–145, 2007.

Groeneveld, D. P., Baugh, W. M., Sanderson, J. S., and Cooper, D. J.: Annual groundwater evapotranspiration mapped from single satellite scenes, J. Hydrol., 344, 146–156, 2007.

Hinojosa-Huerta, O., Nagler, P. L., Carillo-Guererro, Y. K., and Glenn, E. P.: Effect of drought on birds and riparian vegetation in the Colorado River Delta, Mexico, Ecol. Eng., 59, 104–110, 2013.

Jarchow, C. J., Nagler, P. L., Glenn, E. P., Ramírez-Hernández, J., and Rodríguez-Burgueno, J. E.: Evapotranspiration by remote sensing: An analysis of the Colorado River Delta before and after the Minute 319 pulse flow to Mexico, Ecol. Eng., 106, 725–732, 2017a.

Jarchow, C. J., Nagler, P. L., and Glenn, E. P.: Greenup and evapotranspiration following the Minute 319 pulse flow to Mexico: An analysis using Landsat 8 normalized difference vegetation index (NDVI) data, Ecol. Eng., 106, 776–783, 2017b.

Nagler, P. L., Cleverly, J., Glenn, E. Lampkin, D., Huete, A., and Wan, Z.: Predicting riparian evapotranspiration from MODIS vegetation indices and meteorological data, Remote Sens. Environ., 94, 17–30, 2005a.

Nagler, P. L., Scott, R., Westenburg, C., Cleverly, J., Glenn, E., and Huete, A.: Evapotranspiration on western U.S. rivers estimated by the Enhanced Vegetation Index from MODIS and data from eddy covariance and Bowen ratio flux towers, Remote Sens. Environ., 97, 337–351, 2005b.

Nagler, P. L., Hinojosa-Huerta, O., Glenn, E. P., Garcia-Hernandez, J., Romo, R., Curtis, C., Huete, A. R., and Nelson, S. G.: Regeneration of Native Trees in the Presence of Invasive Saltcedar in the Colorado River Delta, Mexico, Conserva.Biol., 19, 1842–1852, 2005c.

Nagler, P. L., Glenn, E. P., Hinojosa-Huerta, O., Zamora, F., and Howard, K.: Riparian Vegetation Dynamics and Evapotranspiration for the Riparian Corridor in the Delta of the Colorado River, Mexico: Implications for Conservation and Management, J. Environ. Manage., 88, 864–874, 2008.

Nagler, P. L., Glenn, E., Scott, R., and Doody, T.: Estimating riparian and agricultural actual evapotranspiration by reference evapotranspiration and MODIS Enhanced Vegetation Index, Special Issue "Advances in Remote Sensing of Crop Water Use Estimation", Remote Sensing, 5, 3849–3871, 2013.

Shanafield, M., Gutierrez Jurado, H., Eliana Rodríguez Burgueño, J., Ramírez Hernández, J., Jarchow, C. J., and Nagler, P. L.: Short- and long-term evapotranspiration rates at ecological restoration sites along a large river receiving rare flow events, Hydrol. Process., 31, 4328–4337, 2017.

Proc. IAHS, 380, 55–66, 2018
https://doi.org/10.5194/piahs-380-55-2018

Open Access
PIAHS

Assessment of SMADI and SWDI agricultural drought indices using remotely sensed root zone soil moisture

Miriam Pablos, Ángel González-Zamora, Nilda Sánchez, and José Martínez-Fernández

Instituto Hispanoluso de Investigaciones Agrarias (CIALE), University of Salamanca (USAL), Villamayor, 37185, Spain

Correspondence: Miriam Pablos (mpablos@usal.es)

Received: 19 April 2018 – Revised: 29 June 2018 – Accepted: 18 July 2018 – Published: 18 December 2018

Abstract. The increasing frequency of drought events has expanded the research interest in drought monitoring. In this regard, remote sensing is a useful tool to globally mapping the agricultural drought. While this type of drought is directly linked to the availability of root zone soil moisture (RZSM) for plants growth, current satellite soil moisture observations only characterize the water content of the surface soil layer (0–5 cm). In this study, two soil moisture-based agricultural drought indices were obtained at a weekly rate from June 2010 to December 2016, using RZSM estimations at 1 km from the Soil Moisture and Ocean Salinity (SMOS) satellite, instead of surface soil moisture (SSM). The RZSM was estimated by applying the Soil Water Index (SWI) model to the SMOS SSM. The Soil Moisture Agricultural Drought Index (SMADI) and the Soil Water Deficit Index (SWDI) were assessed over the Castilla y León region (Spain) at 1 km spatial resolution. They were compared with the Atmospheric Water Deficit (AWD) and the Crop Moisture Index (CMI), both computed at different weather stations distributed over the study area. The level of agreement was analyzed through statistical correlation. Results showed that the use of RZSM does not influence the characterization of drought, both for SMADI and SWDI.

1 Introduction

In the last years, drought has been one of the natural disasters with the worst impact in the agricultural regions worldwide (FAO, 2018). Traditional drought indices, such as the Palmer Drought Severity Index, PDSI (Palmer, 1965), the Standardized Precipitation Index, SPI (McKee et al., 1993) or the Atmospheric Water Deficit, AWD (Purcell et al., 2003), utilize meteorological variables as drought indicators. However, the agricultural drought begins when the available soil moisture drops below a critical threshold, which can cause crop stress and adversely affect yields (Panu and Sharma, 2002; Mishra and Singh, 2010). Then, the soil moisture can be considered as the key variable of this type of drought.

The most used agricultural drought index is the Crop Moisture Index, CMI (Palmer, 1968). It is usually computed alongside the PDSI from evapotranspiration (ET_0) deficit and moisture excess, using climate-based data. Furthermore, several soil moisture-based agricultural drought indices are being developed, as the Soil Moisture Deficit Index, SMDI (Narasimhan and Srinivasan, 2005), the Soil Water Deficit Index, SWDI (Martínez-Fernández et al., 2015) and the Soil Moisture Agricultural Drought Index, SMADI (Sánchez et al., 2016), among others.

Nowadays, there are two missions specifically dedicated to global measuring soil moisture. The first is the Soil Moisture and Ocean Salinity (SMOS), launched in 2009 by the European Space Agency, ESA (Kerr et al., 2016). SMOS L2 surface soil moisture at 15 km is operationally distributed by ESA. In addition, a downscaled SMOS L4 surface soil moisture at 1 km is provided by the Barcelona Expert Centre (BEC). The second satellite is the Soil Moisture Active Passive (SMAP) of the National Aeronautics and Space Administration (NASA), in orbit since 2015 (Chan et al, 2016). SMAP L2 surface soil moisture maps at 36 and 9 km are disseminated by NASA. These remotely sensed soil moisture observations are only capable of measuring the surface soil moisture (SSM) of the top soil layer (approx. 0–5 cm). Meanwhile, the roots of common crops are enclosed at deeper soil layers, around 0–1 m depth (Allen et al., 1998), where the root zone soil moisture (RZSM) is defined.

The RZSM can be estimated using hydrological or land surface models. Commonly, these models are complex, requiring a high computational processing cost and many input variables as well as data assimilation techniques (Muñoz-Sabater et al., 2007; Das et al., 2010; Dumedah et al., 2015). There are two operational RZSM products based on SMOS and SMAP, the SMOS L4 RZSM at 25 km provided by the Centre d'Etudes Spatiales de la Biosphere, CESBIO (Al Bitar et al., 2013) and the SMAP L4 RZSM at 9 km, provided by the National Snow and Ice Data Center Distributed Active Archive Center, NSIDC DAAC (Reichle et al., 2017). An alternative to these hydrological models is the Soil Water Index, SWI (Wagner et al., 1999; Albergel et al., 2008). The SWI model is an exponential filter that is easy to apply for estimating the RZSM.

Since the soil profile behaves as a filter between incoming water and processes that remove it from the hydrological system (Entekhabi et al., 1996), the SSM can be highly affected by anomalous temperature or precipitation events of a particular day. Under these circumstances, the SSM may reflect inaccurate soil water conditions, whereas the RZSM actually indicates the available water storage for plants growth. Therefore, the RZSM is expected to be more appropriate than the SSM for the agricultural drought monitoring. Nonetheless, a limited number of drought indices have been calculated using in situ RZSM (Hunt et al., 2009; Martínez-Fernández et al., 2015), but few drought studies were found using the RZSM estimated from satellite SSM observations (Tobin et al., 2017).

This work aimed to assess the impact of the remotely sensed RZSM in two soil moisture-based agricultural drought indices, the SMADI and the SWDI, instead of using SSM. Both SMADI and SWDI indices were weekly estimated from June 2010 to December 2016 over the agricultural areas of the Castilla y León region, Spain. To assess the behavior of the indices, they were compared through statistical correlation with AWD and CMI, both obtained from meteorological data acquired at different weather stations distributed over the study area.

2 Study area

This research was developed in the Castilla y León region, located at the northwest of Spain (39.85–43.35° N; 7.35–1.65° W), as study area. This region is mainly characterized by a large plain surrounded by mountains to the North, South and East. The extension of the plain is approximately 65 000 km^2 and has a mean altitude of 800 m. Its climate is continental semi-arid Mediterranean with a mean annual temperature of 11.8 °C. The Castilla y León region suffers of scarcity of water, with an average annual precipitation of 450 mm (González-Zamora et al., 2015). Despite this, it is one of the largest agricultural areas of the European Union, chiefly dedicated to rainfed agriculture. In this region, croplands are usually located in areas with an elevation lower than 1100 m, while areas at higher altitudes are mainly covered by forest-pasture. The main crop types are cereals (73.5 %), industrial crops (13 %), forages (8 %), and legumes, tubers and vegetables (5.5 %) (Pablos et al., 2017).

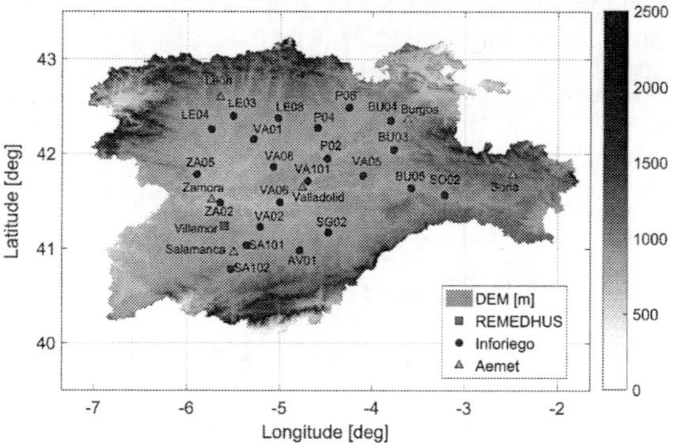

Figure 1. Digital elevation model (DEM) map of the Castilla y León region showing the location of the REMEDHUS, Inforiego and AEMet stations.

3 Data

3.1 In situ data

Three climatic networks were used in this study (Fig. 1). The first is the Soil Moisture Measurement Stations Network of the University of Salamanca, REMEDHUS (González-Zamora et al., 2015). Several weather stations are also installed in REMEDHUS, from which the Villamor station was selected. The second is Inforiego, an agro-meteorological network from the Agriculture Technological Institute of Castilla y León (ITACyL) that is dedicated to irrigation assessment. The third one is the Spanish Meteorological Agency (AEMet) network, which provided long-term weather data.

Daily data (precipitation, air temperature, relative humidity, solar radiation and wind speed) measured by the 22 agro-meteorological stations of the Inforiego network and the Villamor station, were used from June 2010 to December 2016. Similarly, historical records from 1985 to 2016 of daily air temperature and precipitation acquired by six stations of the AEMet network were utilized.

As other ancillary data, the digital elevation model (DEM) provided by the ITACYL at 25 m of spatial resolution was used for discriminating the agricultural areas of the study region.

The surface soil database of the Duero basin from the ITACyL was also employed. It contains approximately ten thousand samples with information of soil texture and organic matter (OM) contents.

3.2 SMOS-BEC L4 SSM

The cloud free SMOS-BEC L4 SSM v.3 product at 1 km over the Iberian Peninsula was used from June 2010 to December 2016. These SSM maps are obtained from the combination of daily SMOS brightness temperature and SSM (L1C and L2, both v.620), 16-day Terra Moderate Resolution Imaging Spectroradiometer (MODIS) Normalized Difference Vegetation Index, NDVI (MOD13A2 v.5) and European Centre for Medium-Range Weather Forecasts (ECMWF) land surface temperature (LST) in a downscaling algorithm (Portal et al., 2017). The downscaling is separately applied for ascending and descending SMOS orbits (06:00 and 18:00 local time, respectively).

3.3 SMOS-CESBIO L4 RZSM

The global SMOS-CESBIO L4 RZSM v.300 at a 25 km Equal Area Scalable Earth (EASE)-2 grid was employed. The RZSM maps are obtained through a double soil layer (0–40 and 40–100 cm) hydrological model (Al Bitar et al., 2013). This model is daily applied for ascending and descending orbits, but it is driven by a 3 day average of the SMOS-CESBIO L3 SSM and ancillary datasets, such as the MODIS NDVI and climatic data.

3.4 SMAP L4 SSM and RZSM

The global SMAP L4 SSM and RZSM v.3 geophysical data maps at a 9 km EASE-2 grid were used from 31 March 2015 to December 2016. Both variables (SSM and RZSM) have a temporal resolution of 3 h and are included in the same product. The SSM and RZSM are obtained by means of the NASA Catchment land surface model. This model is driven by the daily SMAP brightness temperature, previously interpolated to 9 km, together with observation-based surface meteorological data forcing, including precipitation (Reichle et al, 2017).

3.5 MODIS reflectance and LST

The day time Aqua MODIS LST at 1 km (MYD11A1 v.6) and surface reflectance (SR) at 500 m (MYD09GA v.6) in red (∼ 660 nm) and near infrared bands (∼ 860 nm), provided by the U.S. Land Processes Distributed Active Archive Center (LP DAAC, https://lpdaac.usgs.gov, last access: 7 August 2018), were used from June 2010 to December 2016. The SR data were employed to calculate the NDVI. Both LST and SR are projected to a tile-based sinusoidal grid. The four tiles covering the study area (h17v04, h17v05, h18v04 and h18v05) were selected.

4 Methods

4.1 Data pre-processing

The agricultural drought indices were only computed at the agricultural areas of the Castilla y León region. These areas were delimited by an elevation criterion and soils with OM content higher than 4 % were also discarded, because usually these conditions are out of the agricultural domain under Mediterranean conditions. To do so, the DEM was interpolated from the 25 m grid to a regular 1 km grid to match the final resolution that was used in this work (Fig. 1). Then, all input data were clipped with a mask, keeping data in areas with altitudes lower than 1100 m and discarding the rest.

Regarding the soil database from the ITACyL, surface samples over forested areas, defined as those with OM contents higher than 4 %, were discarded. The remaining samples were interpolated to the same 1 km grid used for the indices to obtain four maps (sand, clay, silt and OM).

The SMOS-BEC L4 SSM maps were firstly resampled to the regular 1 km grid. For each day, the ascending and descending maps were averaged. The resulting daily maps were weekly averaged, because the weekly temporal rate fits better with the agricultural management schedule.

In MODIS data, the 4 tiles with the LST and SR maps were mosaicked and clipped to select the study area. Later, the resulting maps were resampled, at their native spatial resolutions (LST at 1 km and SR at 500 m) from the sinusoidal grid to a regular one. The daily SR maps were masked out using the cloud and snow/ice flags, and filtered at the highest quality. The daily NDVI maps at 500 were computed from the SR, using both bands. The NDVI maps at 500 m were then aggregated to the same grid of the LST. Finally, the daily LST and NDVI maps at 1 km were also weekly averaged.

4.2 Estimation of root zone soil moisture

In order to estimate a remotely sensed RZSM, the SWI model (Wagner et al., 1999) was used. The SWI was applied for each day i following a recursive formulation (Albergel et al., 2008):

$$\mathrm{SWI}_i = \mathrm{SWI}_{i-1} + K_i(\mathrm{SSM}_i - \mathrm{SWI}_{i-1}), \quad (1)$$

where SSM_i is the SMOS-BEC L4 SSM at 1 km for the day i, SWI_{i-1} is the SWI (i.e. the RZSM estimation) of the previous day $i-1$, and K_i corresponds to the gain of the day i, which is derived by (Albergel et al., 2008):

$$K_i = \frac{K_{i-1}}{K_{i-1} + e^{-\left(\frac{t_i - t_{i-1}}{T}\right)}}, \quad (2)$$

where K_{i-1} denotes the gain of previous day $i-1$, t_i and t_{i-1} are the times of current day i and previous day $i-1$, respectively, and the T parameter stands for a given characteristic time length. The SWI was initialized using $\mathrm{SWI}_1 = \mathrm{SSM}_1$ and $K_1 = 1$.

The T parameter represents the rate of water transfer of each soil type (Ceballos et al., 2005; Albergel et al., 2008; González-Zamora et al., 2016). The optimal T values were obtained by comparing the SMAP L4 SSM and RZSM. Firstly, the 3 h SSM and RZSM maps were daily averaged. For each pixel, the daily SMAP SSM time series was introduced in the SWI expression with different T values varying from 1 to 100 days, obtaining 100 different possible SWI time series. Then, all the 100 SWI time series were compared to the daily SMAP RZSM time series. The optimal T was that obtaining the highest correlation during this comparison (Ford et al., 2014; González-Zamora et al., 2016). The resulting T map at 9 km EASE-2 grid was interpolated to a regular 1 km grid. Hence, this optimal T together with the daily SMOS-BEC L4 SSM, both at 1 km, were used as inputs of the SWI. Finally, the derived daily SMOS SWI maps were also weekly averaged.

4.3 Estimation of drought indices

4.3.1 Atmospheric Water Deficit (AWD)

The AWD was previously estimated as the difference of evapotranspiration minus precipitation (Purcell et al, 2003). In this study, the weekly AWD was reversely calculated by applying a 7-day running sum separately to both input variables, as in Torres et al. (2013):

$$\mathrm{AWD}_n = P_n - \mathrm{ET}_{0_n}, \tag{3}$$

where P_n is the accumulated precipitation and ET_{0_n} stands for the accumulated evapotranspiration of each week n.

The daily precipitation was measured at the 22 Inforiego stations and at Villamor. The daily evapotranspiration was estimated by the Penman-Monteith method of the Food and Agriculture Organization of the United Nations, FAO (Allen et al., 1998).

4.3.2 Crop Moisture Index (CMI)

The self-calibrating PDSI tool (Wells, 2003) was employed to compute the weekly CMI (Palmer et al., 1968). The required inputs were: (i) the monthly (weekly) averaged air temperature, (ii) the monthly (weekly) accumulated precipitation, (iii) the monthly (weekly) averaged air temperature along the study period and (iv) the total available water (TAW) content. In this research, the CMI was calculated using the daily air temperature and precipitation measured in the six AEMet stations of the Castilla y León region from 1985 to 2016.

The TAW at the root zone was computed as (Allen et al., 1998):

$$\mathrm{TAW} = 1000 Z_\mathrm{r}(\mathrm{FC\text{-}WP}), \tag{4}$$

where Z_r denotes the root depth, and FC and WP are the soil water contents at field capacity and wilting point, respectively.

A root depth of 1 m was considered, as in Pablos et al. (2017). The FC and WP were derived by applying the pedotransfer functions (PTFs) from Rawls et al. (1982) to the surface soil data from the sand, clay and OM content of the pixels at 1 km overlapping the AEMet stations.

4.3.3 Soil Moisture Agricultural Drought Index (SMADI)

The Soil Moisture Condition Index (SMCI), the Modified Temperature Condition Index (MTCI) and the Vegetation Condition Index (VCI) were needed for the SMADI calculation. Both MTCI and VCI were weekly computed as in Sánchez et al. (2016, 2017, 2018):

$$\mathrm{SMCI}_n = \frac{\mathrm{SM}_{\mathrm{max}} - \mathrm{SM}_n}{\mathrm{SM}_{\mathrm{max}} - \mathrm{SM}_{\mathrm{min}}}, \tag{5}$$

$$\mathrm{MTCI}_n = \frac{\mathrm{LST}_n - \mathrm{LST}_{\mathrm{min}}}{\mathrm{LST}_{\mathrm{max}} - \mathrm{LST}_{\mathrm{min}}}, \tag{6}$$

$$\mathrm{VCI}_n = \frac{\mathrm{NDVI}_n - \mathrm{NDVI}_{\mathrm{min}}}{\mathrm{NDVI}_{\mathrm{max}} - \mathrm{NDVI}_{\mathrm{min}}}, \tag{7}$$

where SM_n is the soil moisture, LST_n is the MODIS LST and NDVI_n stands for the MODIS NDVI, all the three variables at 1 km for the current week n. The subscripts max and min correspond to the maximum and minimum weekly values of LST or NDVI for the study period, respectively.

The weekly SMADI was calculated using the aforementioned condition indices (Sánchez et al, 2016, 2017, 2018):

$$\mathrm{SMADI}_n = \mathrm{SMCI}_n \frac{\mathrm{MTCI}_n}{\mathrm{VCI}_{n+2}}, \tag{8}$$

where SMCI_n and MTCI_n are the SMCI and MTCI of the present week n, respectively, and VCI_{n+2} corresponds to the VCI of 2 weeks after. This period of 2 weeks was selected for the time adjustment of the VCI, because it provided the best correlation values when comparing with the other drought indices (Sánchez et al., 2016).

Two versions of the weekly SMCI were computed in this study depending on the soil moisture term (either the SMOS-BEC L4 SSM or the SMOS-derived SWI). Accordingly, two versions of SMADI were weekly computed ($\mathrm{SMADI}_{\mathrm{SSM}}$ and $\mathrm{SMADI}_{\mathrm{SWI}}$, respectively).

4.3.4 Soil Water Deficit Index (SWDI)

The weekly SWDI was computed following Martínez-Fernández et al. (2015, 2016):

$$\mathrm{SWDI}_n = 10\left(\frac{\mathrm{SM}_n - \mathrm{FC}}{\mathrm{AWC}}\right), \tag{9}$$

where the SM_n is the soil moisture of the present week n, FC is the water content at field capacity and AWC corresponds to the available water content for plants growth, which was calculated as the difference between FC and WP.

As for SMADI, two different versions of SWDI were calculated ($SWDI_{SSM}$ and $SWDI_{SWI}$). The first version used the SMOS-BEC L4 SSM and surface soil water parameters. The FC and WP at the surface were computed as equal as in the CMI. The second version employed the SMOS-derived SWI and root zone soil water parameters. The FC and WP at the root zone were estimated through the percentiles method (Martínez-Fernández et al., 2015), which was applied to the current SMOS-CESBIO L4 RZSM product at 25 km. For that, the ascending and descending SMOS-CESBIO L4 RZSM were daily averaged. The 95th and 5th percentiles maps were computed. Then, they were interpolated from 25 km to the regular 1 km grid.

4.4 Comparison strategy

The two versions of SMADI and SWDI were processed and four datasets of weekly maps ($SMADI_{SSM}$, $SMADI_{SWI}$, $SWDI_{SSM}$ and $SWDI_{SWI}$) were estimated over the agricultural areas of the Castilla y León region. The four time series of the pixels at 1 km covering the in situ stations were extracted. They were compared with the AWD time series at the Inforiego and Villamor stations, and the CMI time series at the AEMet stations. Since there is not a precise and universal definition of the drought concept, the AWD and CMI cannot be assumed as absolute "ground-truth", but a benchmark of drought conditions computed from independent ground-based data. In addition, these meteorological-based datasets have been a good proxy for the assessment of agricultural drought in previous research (Martínez-Fernández et al., 2015, 2016; Pablos et al., 2017; Paredes-Trejo and Barbosa, 2017; Sánchez et al., 2016, 2017, 2018). The level of agreement between the time series was analyzed by means of the Pearson's correlation coefficient (R). To assess a possible mismatch of time between drought indices, the $SMADI_{SWI}$ and $SWDI_{SWI}$ were also compared with AWD and CMI of the same week and with those of 1, 2 and 3 antecedent weeks.

5 Results and discussion

5.1 Comparison with AWD

The weekly evolution of the AWD, $SMADI_{SWI}$ and $SWDI_{SWI}$ time series showed very similar seasonal cycles (Fig. 2), alternating dry periods during summer and wet periods during winter. Furthermore, the three agricultural drought indices adequately captured the vegetation growing season of the study area. Since SMADI defines drought with an opposite sign than AWD, CMI and SWDI (Sánchez et al, 2016, 2017, 2018), its vertical axis was plotted increasing downward and its drought threshold (SMADI = 1) was aligned to the zero of the other indices.

There was a certain delay between the $SMADI_{SWI}$ variations with respect to the AWD variations, especially at the onset of the drought events (Fig. 2a). The quicker response of AWD is consistent with the nature of this drought index, directly linked to processes that occur at the surface-atmosphere layer. This delay was also observed in a previous study, where SMADI was estimated from the SMOS SSM (Pablos et al., 2017). In that case, the delay of SMADI was related to the time lag between the occurrence of drought and the NDVI changes (Ji and Peters, 2003; Wang et al., 2001).

A similar delay was also evident between the AWD and $SWDI_{SWI}$, not only at the drought beginning but also at the endings (Fig. 2b). This is intrinsically related to the use of the SWI, because the SSM changes are faster than the water changes of deeper soil layers. The different dynamics of the processes that occur at the atmosphere and the soil systems produced around a one-week lag of the soil moisture variations under meteorological drought conditions (Changnon, 1987). The delay of SWDI was also observed when the SWDI was computed using the in situ RZSM (Martínez-Fernández et al., 2015), but it was not detected when using the SMOS SSM (Martínez-Fernández et al., 2016; Pablos et al., 2017; Paredes-Trejo and Barbosa, 2017). Similar results were obtained at the other Inforiego stations of this work and Villamor (not shown).

The correlations obtained from the comparison of $SMADI_{SSM}$, $SMADI_{SWI}$, $SWDI_{SSM}$ and $SWDI_{SWI}$ with AWD (Fig. 3) showed values that were in line with the results of Fig. 2. As expected, negative correlations were obtained when comparing SMADI and AWD. In both SSM-derived indices, strong correlations were obtained from the comparison with AWD of the same week ($R \approx -0.59$ for $SMADI_{SSM}$ and $R \approx 0.77$ for $SWDI_{SSM}$, in median). These results were similar to those obtained in several previous studies (Martínez-Fernández et al., 2016; Pablos et al., 2017; Paredes-Trejo and Barbosa, 2017). By contrast, the correlations between the SWI-derived indices with AWD of the same week became slightly weaker for $SMADI_{SWI}$ ($R \approx -0.55$, in median, Fig. 3a) and noteworthy weaker for $SWDI_{SWI}$ ($R \approx 0.60$, in median, Fig. 3b).

In both indices, the comparison against the AWD computed 1, 2 or 3 weeks before performed slightly stronger correlations for $SMADI_{SWI}$($R \approx -0.61$, -0.63 and -0.62, respectively, in median) and considerably higher correlations for $SWDI_{SWI}$ ($R \approx 0.72$, 0.71 and 0.71, respectively, in median) than when no time lag was taken into account. A lag of 2–3 weeks of the vegetation response to precipitation was already detected by Zhang et al. (2013).

In general, considering the 25th and 50th percentiles of the boxplots, a delay duration of two weeks obtained the best correlations for both indices. This time lag approximately agrees with the median value of the optimal T over the study area (13 days). Therefore, the SSM-derived indices performed similar degree of agreement with AWD of the same week than that of the SWI-derived indices with AWD of 2 antecedent weeks.

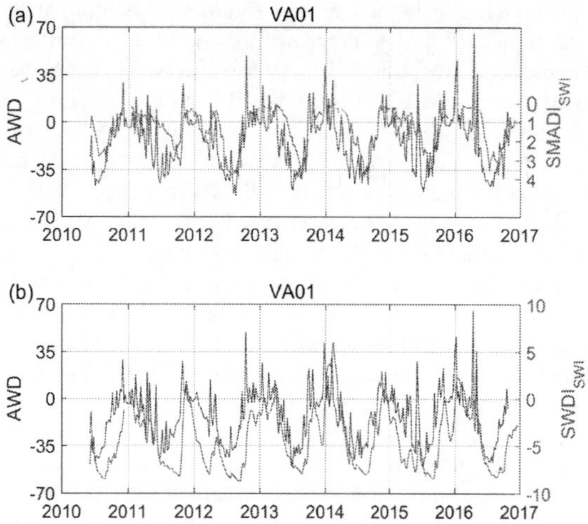

Figure 2. Time series of the weekly AWD and SMADI$_{SWI}$ **(a)**, and SWDI$_{SWI}$ **(b)** at the VA01 station of Inforiego, as an example. Since SMADI = 1 and AWD = 0 are the threshold for drought (Pablos et al., 2017), both values coincide in the *y*-axis. In addition, note that the *y*-axis of SMADI is oriented downward.

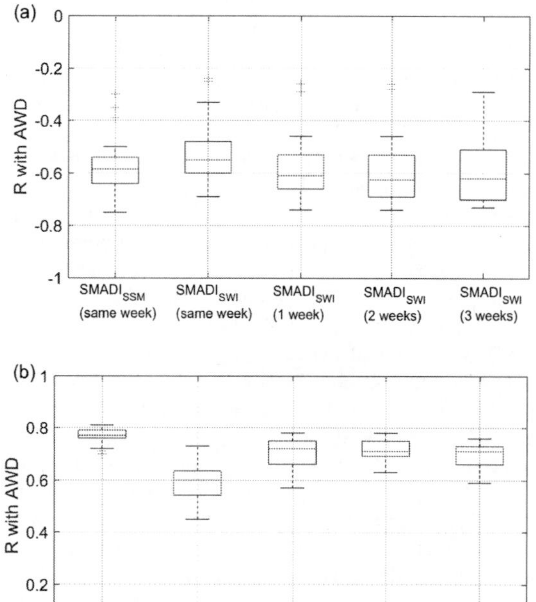

Figure 3. Correlation (R) of the SMADI **(a)** and the SWDI **(b)** obtained using the SMOS SSM and SWI, with the AWD of the same week and 1, 2 or 3 antecedent weeks at the 22 Inforiego stations and Villamor. All values were significant (p-value < 0.05).

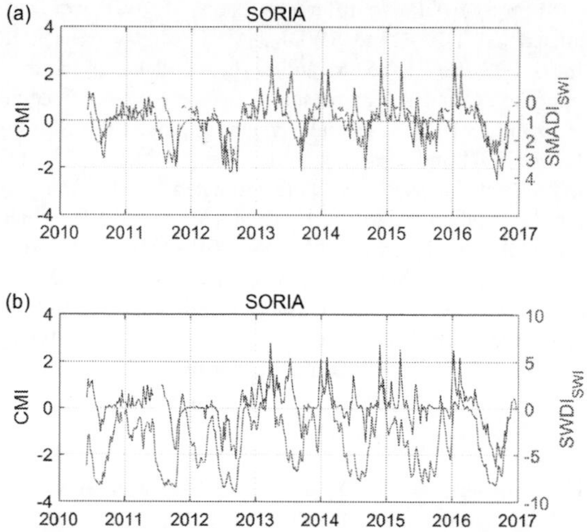

Figure 4. Time series of the weekly CMI and SMADI$_{SWI}$ **(a)**, and SMDI$_{SWI}$ **(b)** at the Soria station of the AEMet, as an example. Since SMADI = 1 and CMI = 0 are the threshold for drought (Pablos et al., 2017), both values coincide in the *y*-axis. In addition, note that the *y*-axis of SMADI is oriented downward.

5.2 Comparison with CMI

The weekly CMI, SMADI$_{SWI}$ and SWDI$_{SWI}$ time series at one AEMet station displayed similar seasonal periods (Fig. 4) than AWD in Fig. 2. Note that SMADI was also plotted with its vertical axis increasing downward. However, no delay was observed between the CMI and the SMADI$_{SWI}$ variations, showing a good correspondence in time (Fig. 4a), as previously found (Pablos et al., 2017; Sánchez et al., 2016, 2018). This could be due to the fact that CMI takes into account the effective soil rooting depth and the soil characteristics, instead of using only climatic data like AWD.

In the case of SWDI$_{SWI}$, there were not conclusive results about the existence of a delay in the comparison with CMI (Fig. 4b). As in the case of AWD, previous research showed a time lag between SWDI and CMI when using the in situ RZSM (Martínez-Fernández et al., 2015), but no delay was found when the SMOS SSM was utilized instead (Martínez-Fernández et al., 2016; Pablos et al., 2017). Similar results were obtained at the other AEMet stations of this work (not shown).

The correlation coefficients obtained from the comparison of SMADI$_{SSM}$, SMADI$_{SWI}$, SWDI$_{SWI}$ and SWDI$_{SWI}$ with CMI (Fig. 5) displayed similar results to those discussed in Fig. 4. As in the case of AWD, negative correlation coefficients were obtained between SMADI and CMI, in agreement with their definitions. Both SMOS SSM-derived and SWI-derived SMADI had very similar correlations in the comparison with CMI of the same week (both $R \approx -0.59$, in median, Fig. 5a). These correlation values decreased in

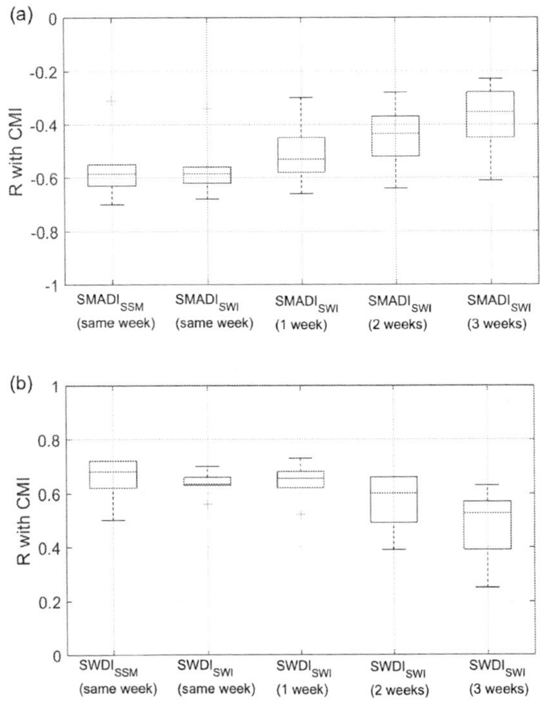

Figure 5. Correlation (R) of the SMADI **(a)** and SWDI **(b)** obtained using the SMOS SSM and SWI, with the CMI of the same week and 1, 2 or 3 antecedent weeks at the six AEMet stations. Only significant values (p-value < 0.05) were considered.

the comparison with CMI of 1, 2, or 3 antecedent weeks ($R \approx -0.54$, -0.44 and -0.36, respectively, in median). This confirms that SMADI and CMI cycles are synchronized, as observed in Fig. 4a.

When analyzing the correlations of CMI with SWDI (Fig. 5b), the use of the SMOS SWI slightly made decrease the correlation coefficients with respect the use of SSM ($R \approx 0.68$ for $\mathrm{SWDI_{SSM}}$ and $R \approx 0.64$ for $\mathrm{SWDI_{SWI}}$, in median). In addition, the correlations between $\mathrm{SWDI_{SWI}}$ and CMI of the antecedent 1, 2 or 3 weeks displayed similar or lower values than when no delay was considered ($R \approx 0.66$, 0.60 and 0.30, in median). This result suggests that there is not lag of time between CMI and $\mathrm{SWDI_{SWI}}$ or, if any, the possible duration would be around one week. Nevertheless, further research is needed to assess this issue as well as to evaluate the results with AWD and CMI in other regions with different environmental conditions. During the comparison with CMI, the level of agreement of the SWI-derived indices was similar to that of the SSM-based indices.

Table 1. Correlation (R) between $\mathrm{SMADI_{SSM}}$ and $\mathrm{SWDI_{SSM}}$, and between $\mathrm{SMADI_{SWI}}$ and $\mathrm{SWDI_{SWI}}$. All values were significant (p-value < 0.05). The number of weeks with available data (N) is also included.

Stations	$\mathrm{SMADI_{SSM}}$ vs $\mathrm{SWDI_{SSM}}$	$\mathrm{SMADI_{SWI}}$ vs $\mathrm{SWDI_{SWI}}$	N (weeks)
AV01	−0.74	−0.80	294
BU03	−0.77	−0.79	286
BU04	−0.75	−0.82	270
BU05	−0.74	−0.80	288
LE03	−0.69	−0.68	303
LE04	−0.28	−0.22	303
LE08	−0.70	−0.74	300
P02	−0.70	−0.77	290
P04	−0.57	−0.65	298
P06	−0.68	−0.74	289
SA101	−0.48	−0.49	272
SA102	−0.49	−0.49	305
SG02	−0.65	−0.71	278
SO02	−0.78	−0.85	288
VA01	−0.74	−0.79	303
VA02	−0.59	−0.69	278
VA05	−0.65	−0.74	274
VA06	−0.63	−0.68	294
VA08	−0.82	−0.87	302
VA101	−0.70	−0.76	290
ZA02	−0.70	−0.68	291
ZA05	−0.65	−0.68	280
VILLAMOR	−0.72	−0.80	276
BURGOS	−0.66	−0.72	239
LEÓN	−0.68	−0.71	303
SALAMANCA	−0.71	−0.75	291
SORIA	−0.76	−0.84	270
VALLADOLID	−0.36	−0.41	287
ZAMORA	−0.81	−0.84	287

5.3 Comparison of SMADI with SWDI

The correlation analysis between SMADI and SWDI clearly showed a high agreement (Table 1), both with SMOS SSM ($R \approx -0.28$ to -0.82; -0.70 in median) and SWI ($R \approx -0.22$ to -0.85; -0.74 in median). Additionally, the correlation differences between the SSM and SWI-derived indices were very low ($\Delta R \approx 0$ to 0.10, in absolute value) to be significant. Thus, the results of both approaches were comparable, as previously observed in the assessment with AWD and CMI. Notwithstanding, in a detailed analysis of the correlation coefficients, 24 out of the total 29 stations obtained stronger correlations in the SWI approach. This suggests a certain trend toward a correlation improvement when the RZSM estimation was used. The high number of weeks with available data ($N \approx 239$ to 305 weeks) of a total 342 weeks ensures that the values were robustly computed, as in the previous comparisons with AWD and CMI. However, no conclusions should be inferred from these results because they may hide self-adjustments or other artifacts behind this direct comparison.

6 Conclusions

Since the launch of recent satellite missions dedicated to soil moisture, new applications have been emerged using it as a key agricultural and hydrological variable. Among them, the number of drought approaches is increasing and the results improving. As a great leap forward, several research tries to overcome the drawback of using the soil moisture at surface level derived from remote measurements. With the same aim, two agricultural drought indices, SMADI and SWDI, were applied using remotely sensed soil moisture both at the surface layer (directly retrieved from the sensor observations) and at the root zone (estimated by applying the SWI model to the SSM). The results showed that the use of the RZSM estimation does not influence the characterization of drought through SMADI and SWDI, both in comparison with a meteorological drought index (AWD) and an agricultural one (CMI).

Both SMADI and SWDI showed a similar capability for agricultural drought monitoring, and it was highlighted that a certain time lag should be computed between the at-surface variables (SSM, precipitation) and the RZSM, due to the different response time of the associated processes. This lag is consistent with the SWI model, which takes into account the time lapse of the water transfer from the surface to the deeper soil layers.

The use of RZSM estimations from remote sensing offers a new opportunity for drought monitoring and, in a broadly sense, for many agricultural management applications.

Data availability. The data of the Villamor station are available upon request. The data of Inforiego are freely accessible (http://www.inforiego.org, ITACyL, 2018a). The data of AEMet are also accessible for free (http://www.aemet.es, AEMet, 2018). The digital elevation model (DEM) of the study area was provided by the ITACyL, as well as the surface soil database (http://suelos.itacyl.es, ITACyL, 2018b). The authors especially thank Gerard Portal and Mercè Vall-llossera from the Technical University of Catalonia (UPC) and the BEC (http://bec.icm.csic.es/land-datasets, BEC, 2018) for providing the new cloud-free SMOS-BEC L4 SSM v.3 product. The SMOS-CESBIO L4 RZSM was generated by the CESBIO and disseminated by the CATDS (http://www.catds.fr/Products/Available-products-from-CEC-SM/L4-Land-research-products, CESBIO and CATDS, 2018). An updated version of the SMAP L4 SSM and RZSM variables are accessible at the NSIDC DAAC (https://nsidc.org/data/SPL4SMGP/versions/4, NSIDC DACC, 2018). The Aqua MODIS LST and surface reflectance were provided by the NASA Land Processes Distributed Active Archive Center (https://lpdaac.usgs.gov, LP DAAC, 2018).

Appendix A: Acronyms

For easy reading of this research, all the acronyms were alphabetically summarized in the following list:

AEMet: Spanish Meteorological Agency

AWD: Available Water Content

AWD: Atmospheric Water Deficit

BEC: Barcelona Expert Centre

CESBIO: Centre d'Etudes Spatiales de la Biosphere

CMI: Crop Moisture Index

DEM: Digital Elevation Model

EASE: Equal Area Scalable Earth

ECMWF: European Centre for Medium-Range Weather Forecasts

ESA: European Space Agency

FC: Field Capacity

ITACyL: Agriculture Technological Institute of Castilla y León

LP DAAC: Land Processes Distributed Active Archive Center

LST: Land Surface Temperature

MODIS: Moderate Resolution Imaging Spectroradiometer

MTCI: Modified Temperature Condition Index

NASA: National Aeronautics and Space Administration

NDVI: Normalized Difference Vegetation Index

NSIDC DAAC: National Snow and Ice Data Center Distributed Active Archive Center

PDSI: Palmer Drought Severity Index

PTF: Pedotransfer Function

REMEDHUS: Soil Moisture Measurements Station Network of the University of Salamanca

RZSM: Root Zone Soil Moisture

SMADI: Soil Moisture Agricultural Drought Index

SMAP: Soil Moisture Active Passive

SMCI: Soil Moisture Condition Index

SMDI: Soil Moisture Deficit Index

SMOS: Soil Moisture and Ocean Salinity

SPI: Standardized Precipitation Index

SSM: Surface Soil Moisture

SWDI: Soil Water Deficit Index

SWI: Soil Water Index

TAW: Total Available Water

VCI: Vegetation Condition Index

WP: Wilting Point

Author contributions. The initial idea for this research was conceived by all the four authors. The REMEDHUS, Inforiego and AEMet climatic data were prepared by ÁGZ. All the satellite data were downloaded by MP. The soil database data and the digital elevation model were prepared by NS. The SMOS SWI was computed by ÁGZ. All the other data processing was performed by MP, who also collected all the results. MP, NS and JMF equally contributed to the interpretation of the results. The first manuscript was written by MP. The four authors revised the final manuscript and approved it.

Competing interests. The authors declare that they have no conflict of interest.

Special issue statement. This article is part of the special issue "Earth Observation for Integrated Water and Basin Management: New possibilities and challenges for adaptation to a changing environment". It is a result of The Remote Sensing & Hydrology Symposium, Cordoba, Spain, 8–10 May 2018.

Acknowledgements. The study was supported by the Junta de Castilla y Léon (JCyL) and the European Regional Development Fund (ERDF) through the project SA007U16, and Spanish Ministry of Economy and Competitiveness with the projects PROMISES: ESP2015-67549-C3-3-R and L-BAND: ESP2017-89463-C3-3-R.

Edited by: Ana Andreu
Reviewed by: two anonymous referees

References

Albergel, C., Rüdiger, C., Pellarin, T., Calvet, J.-C., Fritz, N., Froissard, F., Suquia, D., Petitpa, A., Piguet, B., and Martin, E.: From near-surface to root-zone soil moisture using an exponential filter: an assessment of the method based on in-situ observations and model simulations, Hydrol. Earth Syst. Sci., 12, 1323–1337, https://doi.org/10.5194/hess-12-1323-2008, 2008.

Al Bitar, A., Kerr, Y. H., Merlin, O., Cabot, F., and Wigneron, J. P.: Global drought index from SMOS soil moisture, IEEE Int. Geos. Remote Sens. Symp. (IGARSS), Melbourne, Australia, 2013.

Allen, R. G., Pereira, L. S., Raes, D., and Smith, M.: Crop evapotranspiration: Guidelines for computing crop water requirements, Food and Agric. Org. of the United Nations (FAO), Rome, Italy, 1998.

AEMet (Spanish Meteorological Agency): Climatic data of the AEMet network, available at: http://www.aemet.es, last access: 7 August 2018.

BEC (Barcelona Expert Centre): SMOS-BEC L4 surface soil moisture product, available at: http://bec.icm.csic.es/land-datasets, last access: 7 August 2018.

CESBIO (Centre d'Etudes Spatiales de la Biosphere) and CATDS (Centre Aval de Traitement des Donées SMOS): SMOS-CESBIO L4 root zone soil moisture product, available at: http://www.catds.fr/Products/Available-products-from-CEC-SM/L4-Land-research-products, last access: 7 August 2018.

Changnon, S: Detecting Drought Conditions in Illinois, Illinois State Water Survey, Champaign, Circular 169, 1987.

Chan, S. K., Bindlish, R., O' Neill, P. E., Njoku, E., Jackson, T. J., Colliander, A., Chen, F., Burgin, M., Dunbar, S., Piepmeier, J., Yueh, S., Entekhabi, D., Cosh, M. H., Caldwell, T., Walker, J., Wu, X., Berg, A., Rowlandson, T., Pacheco, A., McNairn, H., Thibeault, M., Martínez-Fernández, J., González-Zamora, Á., Seyfried, M., Bosch, D., Starks, P., Goodrich, D., Prueger, J., Palecki, M., Small, E. E., Zreda, M., Calvet, J. C., Crow, W. T., and Kerr, Y. H.: Assessment of the SMAP passive soil moisture product, IEEE T. Geosci. Remote, 54, 4994–5007, https://doi.org/10.1109/tgrs.2016.2561938, 2016.

Ceballos, A., Scipal, K., Wagner, W., and Martínez-Fernández, J.: Validation of ERS scatterometer-derived soil moisture data in the central part of the Duero Basin, Spain, Hydrol. Process., 19, 1549–1566, https://doi.org/10.1002/hyp.5585, 2005.

Das, N. N., Mohanty, B. P., and Njoku, E. G.: Profile soil moisture across spatial scales under different hydroclimatic conditions, Soil Sci., 175, 315–319, https://doi.org/10.1097/SS.0b013e3181e83dd3, 2010.

Dumedah, G., Walker, J. P., and Merlin, O.: Root-zone soil moisture estimation from assimilation of downscaled soil moisture and ocean salinity data, Adv. Water Resour., 84, 14–22, https://doi.org/10.1016/j.advwatres.2015.07.021, 2015.

Entekhabi, D., Rodriguez-Iturbe, I., and Castelli, F.: Mutual interaction of soil moisture state and atmospheric processes, J. Hydrol., 184, 3–17, https://doi.org/10.1016/0022-1694(95)02965-6, 1996.

FAO: 2017 The impact of disasters and crises in agriculture and food security, Food and Agriculture Organization of the United Nations (FAO), Italy, 2018.

Ford, T. W., Harris, E., and Quiring, S. M.: Estimating root zone soil moisture using near-surface observations from SMOS, Hydrol. Earth Syst. Sci., 18, 139–154, https://doi.org/10.5194/hess-18-139-2014, 2014.

González-Zamora, Á., Sánchez, N., Martínez-Fernández, J., Gumuzzio, A., Piles, M., and Olmedo, E.: Long-term smos soil moisture products: A comprehensive evaluation across scales and methods in the Duero basin (Spain), Phys. Chem. Earth, 83–84, 123–136, https://doi.org/10.1016/j.pce.2015.05.009, 2015.

González-Zamora, Á., Sánchez, N., Martínez-Fernández, J., and Wagner, W.: Root-zone plant available water estimation using the SMOS-derived soil water index, Adv. Water Resour., 96, 339–353, https://doi.org/10.1016/j.advwatres.2016.08.001, 2016.

Hunt, E. D., Hubbard, K. G., Wilhite, D. A., Arkebauer, T. J., and Dutcher, A. L.: The development and evaluation of a soil moisture index, Int. J. Climatol., 29, 747–759, https://doi.org/10.1002/joc.1749, 2009.

ITACyL (Agriculture Technological Institute of Castilla y León): Climatic data of the Inforiego network, available at: http://www.inforiego.org, last access: 7 August 2018a.

ITACyL (Agriculture Technological Institute of Castilla y León): Surface soil database, available at: http://suelos.itacyl.es, last access: 7 August 2018b.

Ji, L. and Peters, A. J.: Assessing vegetation response to drought in the northern great plains using vegetation and drought indices, Remote Sens. Environ., 87, 85–98, https://doi.org/10.1016/S0034-4257(03)00174-3, 2003.

Kerr, Y. H., Al-Yaari, A., Rodríguez-Fernández, N., Parrens, M., Molero, B., Leroux, D., Bircher, S., Mahmoodi, A., Mialon, A., Richaume, P., Delwart, S., Al Bitar, A., Pellarin, T., Bindlish, R., Jackson, T. J., Rüdiger, C., Waldteufel, P., Mecklenburg, S., and Wigneron, J. P.: Overview of SMOS performance in terms of global soil moisture monitoring after six years in operation, Remote Sens. Environ., 180, 40–63, https://doi.org/10.1016/j.rse.2016.02.042, 2016.

LP DAAC (Land Processes Distributed Active Archive Center): MODIS data, https://lpdaac.usgs.gov, last access: 7 August 2018.

Martínez-Fernández, J., González-Zamora, A., Sánchez, N., and Gumuzzio, A.: A soil water based index as a suitable agricultural drought indicator, J. Hydrol., 522, 265–273, https://doi.org/10.1016/j.jhydrol.2014.12.051, 2015.

Martínez-Fernández, J., González-Zamora, Á., Sánchez, N., Gumuzzio, A., and Herrero-Jiménez, C. M.: Satellite soil moisture for agricultural drought monitoring: Assessment of the smos derived soil water deficit index, Remote Sens. Environ., 177, 277–286, https://doi.org/10.1016/j.rse.2016.02.064, 2016.

McKee, T. B., Doesken, N. J., and Kleist, J.: The relationship of drought frequency and duration to time scales, Proceedings of the 8th Conference on Applied Climatology Boston, MA, 179–183, 1993.

Mishra, A. K. and Singh, V. P.: A review of drought concepts, J. Hydrol., 391, 202–216, https://doi.org/10.1016/j.jhydrol.2010.07.012, 2010.

Muñoz-Sabater, J., Jarlan, L., Calvet, J. C., Bouyssel, F., and De Rosnay, P.: From near-surface to root-zone soil moisture using different assimilation techniques, J. Hydrometeorol., 8, 194–206, https://doi.org/10.1175/jhm571.1, 2007.

Narasimhan, B. and Srinivasan, R.: Development and evaluation of soil moisture deficit index (SMDI) and evapotranspiration deficit index (ETDI) for agricultural drought monitoring, Agr. Forest Meteorol., 133, 69–88, https://doi.org/10.1016/j.agrformet.2005.07.012, 2005.

NSIDC DACC (National Snow and Ice Data Center Distributed Active Archive Center),:SMAP L4 soil moisture product, available at: https://nsidc.org/data/SPL4SMGP/versions/4, last access: 7 August 2018.

Pablos, M., Martínez-Fernández, J., Sánchez, N., and González-Zamora, Á.: Temporal and Spatial Comparison of Agricultural Drought Indices from Moderate Resolution Satellite Soil Moisture Data over Northwest Spain, Remote Sens., 9, 1168, https://doi.org/10.3390/rs9111168, 2017.

Palmer, W. C.: Meteorological drought, in: Research paper no. 45, edited by: Connor, J. T. and White, R. M., U.S. Weather Bureau, Washington, DC, USA, 1965.

Palmer, W. C.: Keeping track of crop moisture conditions, nationwide: The new crop moisture index, Weatherwise, 21, 156–161, https://doi.org/10.1080/00431672.1968.9932814, 1968.

Panu, U. S. and Sharma, T.: Challenges in drought research: Some perspectives and future directions, Hydrol. Sci. J., 47, S19–S30, https://doi.org/10.1080/02626660209493019, 2002.

Paredes-Trejo, F. and Barbosa, H.: Evaluation of the SMOS-derived Soil Water Deficit Index as agricultural drought index in northeast of Brazil, Water, 9, 377, https://doi.org/10.3390/w9060377, 2017.

Portal, G., Vall-llossera, M., Piles, M., Camps, A., Chaparro, D., Pablos, M., and Rossato, L.: A spatially consistent downscaling approach for SMOS using an adaptive moving window, IEEE Int. Geosc. Remote Sens. Symp. (IGARSS) 4151–4153, https://doi.org/10.1109/IGARSS.2017.8127915, Fort Worth, Texas, USA, 2017.

Purcell, L. C., Sinclair, T. R., and McNew, R. W.: Drought avoidance assessment for summer annual crops using long-term weather data research supported in part by the united soybean board, project no. 1238, Agron. J., 95, 1566–1576, https://doi.org/10.2134/agronj2003.1566, 2003.

Rawls, W. J., Brakensiek, D. L., and Saxtonn, K. E.: Estimation of soil water properties, T. ASABE, 25, 1316–1320, https://doi.org/10.13031/2013.33720, 1982.

Reichle, R. H., De Lannoy, G. J. M., Liu, Q., Ardizzone, J. V., Colliander, A., Conaty, A., Crow, W., Jackson, T. J., Jones, L. A., Kimball, J. S., Koster, R. D., Mahanama, S. P., Smith, E. B., Berg, A., Bircher, S., Bosch, D., Caldwell, T. G., Cosh, M., González-Zamora, A., Collins, C. D. H., Jensen, K. H., Livingston, S., Lopez-Baeza, E., Martínez-Fernández, J., McNairn, H., Moghaddam, M., Pacheco, A., Pellarin, T., Prueger, J., Rowlandson, T., Seyfried, M., Starks, P., Su, Z., Thibeault, M., van der Velde, R., Walker, J., Wu, X., and Zeng, Y.: Assessment of the SMAP level-4 surface and root-zone soil moisture product using in situ measurements, J. Hydrometeorol., 18, 2621–2645, https://doi.org/10.1175/jhm-d-17-0063.1, 2017.

Sánchez, N., González-Zamora, Á., Piles, M., and Martínez-Fernández, J.: A new soil moisture agricultural drought index (SMADI) integrating MODIS and SMOS products: A case of study over the Iberian Peninsula, Remote Sens., 8, 287, https://doi.org/10.3390/rs8040287, 2016.

Sánchez, N., González-Zamora, Á., Martínez-Fernández, J., Piles, M., Pablos, M., Wardlow, B., Tadesse, T., and Svoboda, M. D.: Preliminary assessment of an integrated smos and modis application for global agricultural drought monitoring, IEEE Int. Geosc. Remote Sens. Symp. (IGARSS), https://doi.org/10.1109/IGARSS.2017.8127374, Fort Worth, Texas, USA, 2017.

Sánchez, N., González-Zamora, Á., Martínez-Fernández, J., Piles, M., and Pablos, M.: Integrated remote sensing approach to global agricultural drought monitoring, Agr. Forest Meteorol., 259, 141–153, https://doi.org/10.1016/j.agrformet.2018.04.022, 2018.

Tobin, K. J., Torres, R., Crow, W. T., and Bennett, M. E.: Multidecadal analysis of root-zone soil moisture applying the exponential filter across CONUS, Hydrol. Earth Syst. Sci., 21, 4403–4417, https://doi.org/10.5194/hess-21-4403-2017, 2017.

Torres, G. M., Lollato, R. P., and Ochsner, T. E.: Comparison of drought probability assessments based on atmospheric water deficit and soil water deficit, Agron. J., 105, 428–436, https://doi.org/10.2134/agronj2012.0295, 2013.

Wang, J., Price, K. P., and Rich, P. M.: Spatial patterns of NDVI in response to precipitation and temperature in the central Great Plains, Int. J. Remote Sens., 22, 3827–3844, https://doi.org/10.1080/01431160010007033, 2001.

Wagner, W., Lemoine, G., and Rott, H.: A method for estimating soil moisture from ers scatterometer and soil data, Remote Sens. Environ., 70, 191–207, https://doi.org/10.1016/S0034-4257(99)00036-X, 1999.

Wells, N., Goddard, S., and Hayes, M. J.: A self-calibrating palmer drought severity index, J. Climate, 17, 2335–2351, https://doi.org/10.1175/1520-0442(2004)017<2335:aspdsi>2.0.co;2, 2004.

Zhang, F., Zhang, L., Wang, X., and Hung, J.: Detecting agro-droughts in southwest of china using MODIS satellite data, J. Integr. Agr., 12, 159–168, https://doi.org/10.1016/S2095-3119(13)60216-6, 2013.

Proc. IAHS, 380, 67–72, 2018
https://doi.org/10.5194/piahs-380-67-2018

Earth Observation for Integrated Water and Basin Management: Challenges for adaptation to a changing environment

Validating improved-MODIS products from spectral mixture-Landsat snow cover maps in a mountain region in southern Spain

Rafael Pimentel[1,3], Carlo Marín[2], Ludovica De Gregorio[2], Mattia Callegari[2], María J. Pérez-Palazón[3], Claudia Notarnicola[2], and María J. Polo[3]

[1]Hydrological Research Unit, Swedish Meteorological and Hydrological Institute, Norrköping, 60176, Sweden
[2]Institute for Earth Observation, EURAC, Bolzano, 39100, Italy
[3]Fluvial Dynamic and Hydrology Research Group, Andalusian Institute for Earth System Research, University of Cordoba, Córdoba, 14071, Spain

Correspondence: Rafael Pimentel (rpimentel@uco.es)

Received: 24 April 2018 – Revised: 15 August 2018 – Accepted: 4 September 2018 – Published: 18 December 2018

Abstract. Remote sensing is the only feasible data source for distributed modelling of snow in mountain regions on medium to large scales, due to the limited access to these areas together with the lack of dense ground monitoring stations for snow variables. Observations worldwide identify snow cover persistence together with snowfall occurrence as the most affected variables by global warming. In Mediterranean regions, the spatiotemporal evolution of the snow cover can experiment quick changes that result in different accumulation-ablation cycles during the cold season. High frequency sensors are required to adequately monitor such shifts; however, for trend analyses, the Landsat time series constitute the only available source of data, being their frequency low for this regime, especially when cloudy conditions limit the available images. On the other hand, the MODIS daily series provide more than 15 years of continuous snow maps, despite the spatial resolution may pose a constraint in areas with abrupt topography; several approaches have been done to improve their spatial resolution from combining different information. This work presents a methodological approach to validate the improved MODIS daily snow cover maps from Notarnicola et al. (2013a, b), with 250 m spatial resolution, in Sierra Nevada (southern Spain), from a reference data set obtained by spectral mixture analyses of Landsat TM data by Pimentel et al. (2017b). This reference time series of fractional snow maps, with 30 m spatial resolution, were validated from high resolution local time series of snow maps obtained by terrestrial time-lapse cameras. The results show a significantly high correlation between the two snow map products both on a global and basin scales in the Sierra Nevada area. Selected areas and time periods are shown to address the convergence and divergence between both products and assess the development of a fusion algorithm to retrieve daily Landsat-resolution snow maps on a long term basis.

1 Introduction

Remote sensing techniques constitute the best source to provide distributed information about the snowpack evolution on medium to long time scales, complementing the traditional in situ field surveys and automatic ground measurements. Snow cover fraction (SCF) is one of the more reliable snow related variable measured from the space (Dozier and Painter, 2004) and is commonly used in hydrological studies to calibrate, evaluate, or be assimilated into snow distributed modelling (Andreadis and Lettenmaier, 2006; Parajka and Blöschl, 2008; Pimentel et al., 2015) Within the different Earth Observation (EO) missions, (1) Landsat-5 (TM), Landsat-7 (ETM+) and Landsat-8 (OLI), with 30×30 m spatial resolution and 16 days revisiting time (Roy et al., 2014; Pimentel et al., 2017a), and (2) MODIS Terra and Aqua, with

Published by Copernicus Publications on behalf of the International Association of Hydrological Sciences.

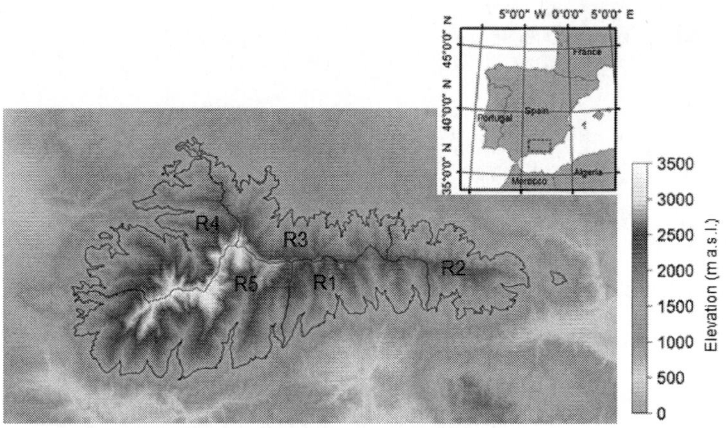

Figure 1. Area of Sierra Nevada Mountain (Spain) above 1500 m a.s.l. and limits of the five regions in which the study area has been divided for the spatial analysis: R1 – Adra, R2 – Andarax, R3 – Fardes, R4 – Genil and R5 – Guadalfeo.

500 m grid cell resolution and daily temporal frequency for snow, are the most extended data sources for snow studies, since they offer the highest spatial and temporal resolution, respectively (Hall et al., 2002).

However, Mediterranean mountainous areas are extremely vulnerable to climate change effects and highly dependent on snow water resources (Barnett et al., 2005; Giorgi, 2006). The particularities of the snowpack make the use of raw EO products not enough to capture these specific patterns. For instance, the very strong spatiotemporal variability, which very often undergoes different accumulation-snowmelt cycles during the cold season in a given year, or the snow patched distribution around local singularities, such as rocks and vegetation, consequence of a very complex ablation process (Ménard et al., 2014; Pimentel et al., 2015, 2017b). Hence, both high temporal and spatial resolutions are required to have a realistic representation of the snow cover.

In this context Pimentel et al. (2017a) carried out a spectral mixture analysis to derive fractional snow cover map time series from Landsat TM and ETM+. High resolution terrestrial photography (TP) was used as ground truth to validate the obtained product. The spatial resolution of the snow cover area was improved using this technique; however, the large revisiting time of Landsat TM and ETM+ in addition to the presence of clouds in some of the dates with snow presence, constitute a big constraint for useful time series. Using the same idea of improving spatial representation of the snow, Notarnicola et al. (2013a, b) developed an algorithm that combines different MODIS products: MOD09GQ-MYD09GQ, MOD09GA-MYD09GA, MOD021KM-MYD021KM, MOD03-MYD03, to produce snow cover maps at 250 m spatial resolution and daily frequency. The algorithm has specific modules to take into account the effect of vegetation and clouds and increase the spatial resolution of the standard snow MODIS product, MOD10A1-MOD10A2, from 500 to 250 m.

This work presents a methodological approach to assess the improved MODIS daily snow cover maps from Notarnicola et al. (2013a, b), in Sierra Nevada (southern Spain), using as reference data set the Landsat fractional cover maps obtained by spectral mixture analyses by Pimentel et al. (2017a).

2 Study site and data available

This study is carried out in Sierra Nevada Mountains, southern Spain (Fig. 1). They are a linear mountain range of 90-km length that runs parallel to the coastline of Mediterranean Sea. Alpine and Mediterranean climate conditions coexist in just a 40-km distance. Strong altitudinal gradients with marked differences between the south (directly affected to the sea) and the north faces are found in the area.

The snow usually appears above 2000 m a.s.l. during winter and spring even though the major snowmelt season generally lasts from April to June, but can be also found at lower altitudes every year. The typically mild Mediterranean winters produce several snowmelt cycles before the final melting phase, which distributes the snow in patches over the terrain. Precipitation and temperature regimes are highly variable among years, with annual precipitation values averaged in the area that can range from 200 to 900 mm and annual mean of the daily minimum and maximum temperature of −5 and 30 °C, respectively (Pérez-Palazón et al., 2015).

Two snow cover EO products are used in this study: (1) Fractional snow cover maps, at 30 × 30 m and 16 days spatial and temporal resolution respectively, derived from spectral mixture analysis of Landsat TM and ETM+ validated using as high resolution terrestrial photography (Pimentel et al., 2017a); (2) binary snow cover maps obtained from an algorithm that combines several MODIS products and produce daily snow cover maps with a grid cell size of 250 m (Notarnicola et al., 2013a, b). In the text these products will be referred as Landsat-mix and MODIS-EURAC, respectively. Both products have a temporal overlapping from 1 January 2002 to 31 August 2013. For this period, a total number of 108 and 2963 cloud-free images were used in the study for Landasat-mix and MODIS-EURAC, respectively; considering as cloudy images those whose presence of clouds exceeded a 10 % of the study area.

3 Methods

SCF was calculated over the whole study area in Sierra Nevada (area above 1500 m a.s.l) and in each one of the five main headwaters regions: R1 – Adra, R2 – Andarax, R3 – Fardes, R4 – Genil and R5 – Guadalfeo, for both snow products, Landsat-mix and MODIS-EURAC.

SCF from both products were compared in the 108 common dates. A simple linear model was fitted in those days to relate both products. Landsat-mix was chosen as ground truth and MODIS-EURAC as dependent variable. Equation (1) was used for that,

$$SCF_{\text{Landsat-mix}} = (SCF_{\text{MODIS-EURAC}} \times a) + b, \quad (1)$$

where a and b are the two parameters of the lineal model. Using this model, the SCF from MODIS-EURAC was reconstructed using the 2963 cloud-free images.

4 Results

Figure 2 shows the evolution of SCF from MODIS-EURAC and Landsat-mix in the overlapping dates for both products in each of the defined regions and over the whole study area. SCF follows the same trend for both products, with a clear overestimation of the SCF derived from MODIS-EURAC. This overestimation is especially significant during the dates with higher SCF values and specifically in R1 – Adra, where differences about 0.20 $m^2\,m^{-2}$ can be found in the 2 days with more snow throughout the study period. Differences are practically negligible during dates with low SCF for all the regions. However, this general overestimation trend from MODIS-EURAC change during the last stages of the snowmelt season, when its lower resolution is not able to capture snow remaining isolated snow patches.

Figure 3 and Table 1 show the linear relation found between the two products and the parameters that fitted these relationships respectively. The linear pattern is clear for all regions, with determination coefficients ranging from 0.979 for R4 – Genil to 0.995 for R2 – Andarax, with a clear overestimation of MODIS-EURAC.

The parameter a (Table 1), which measures the slope of the fitted model and consequently determines the magnitude

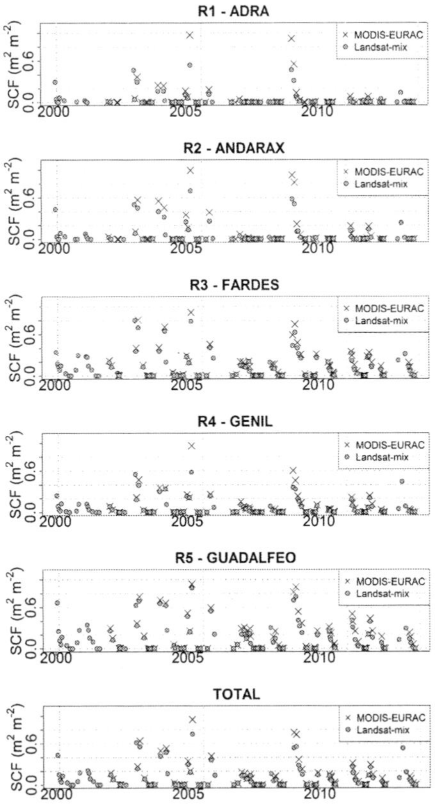

Figure 2. Comparison between SCF evolution of both products MODIS-EURAC (black crosses) and Landsat-mix (gray dots) in the 108 overlapping dates, in each of the regions selected and in the whole study area.

of the MODIS-EURAC overestimation, differs between regions. Lower values of the parameters, which imply higher overestimations, are found in the drier and warmer areas (Pérez-Palazón et al., 2015), 0.700 in R1 – Adra and 0.645 in R2 – Andarax; located in the south face and with lower mean elevation. On the contrary, wetter and colder regions have higher values and consequently less overestimation coming from MODIS-EURAC. Although, the general accuracy from MODIS snow products is estimated approximately at 93 % (Hall and Riggs, 2007) and similar studies has found an accuracy of 94.6 % comparing MODIS products with surface observations over northern China, (Huang et al., 2016), the heterogeneity of the snow distribution due to the abrupt terrain and climate conditions, make the overestimations of MODIS-ERURAC over this are slightly bigger.

The clear linear fit found allows using this relationship as a simple model to correct the average values calculated using MODIS-EURAC over the study area. Overestimation corrections of 0.30, 0.35, 0.13, 0.20 and 0.23 $m^2\,m^{-2}$, were achieved for Adra, Andarax, Fardes, Genil and Guadalfeo,

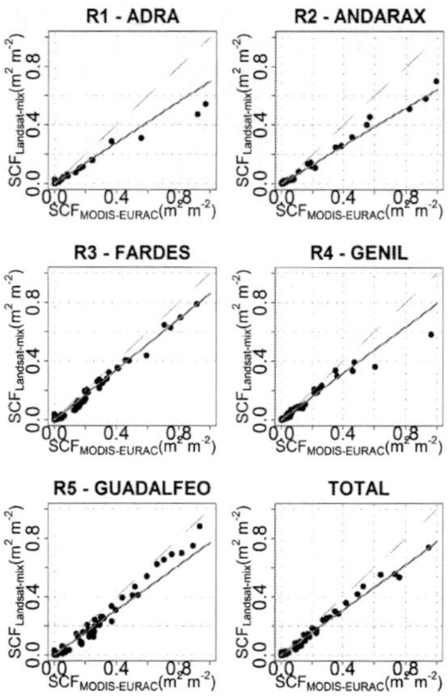

Figure 3. Dispersion graphs comparing both products MODIS-EURAC (x-axis) and Landsat-mix (y-axis) in the 108 selected dates along the study period. Linear fit defined following Eq. (1) and the parameters of Table 1 (red line). 1 : 1 line in light grey.

Table 1. Fitted parameters and coefficient of determination (a, b and R^2) of the linear model that related both products, MODIS-EURAC and Landsat-mix in each of the region and in the whole study area.

	a	b	R^2
R1	0.700	0	0.992
R2	0.645	0	0.995
R3	0.862	0	0.993
R4	0.792	0	0.979
R5	0.770	0	0.988
Total	0.775	0	0.993

respectively; with a mean value of 0.23 $m^2\,m^{-2}$ for the whole study area. Figure 4 shows an example of the new corrected values of MODIS-EURAC for the hydrological year 2004–2005. The general MODIS-EURAC overestimation has been reduced giving more realistic values of the total SCF over the study area.

Tables 2 and 3 show the annual mean and maximum SCF calculated for the three snow products, Landsat-mix (Land), MODIS-EURAC (MOD) and Corrected MODIS-EURAC (NEW) in each of the study region and the whole study area, respectively. Both tables show the clear impact that this simple correction has on the quality of the results. The corrected MODIS-EURAC combines the spatial accuracy of the Landsat-mix and the high temporal resolution of the MODIS-EURAC products. Moreover, the direct use of

Table 2. Annual mean average snow cover fraction (SCF) calculated for the three snow products, Landsat-mix (Land), MODIS-EURAC (MOD) and Corrected MODIS-EURAC (NEW) in each of the study region and the whole study area.

	Region 1			Region 2			Region 3			Region 4			Region 5			Total		
	Land	MOD	NEW	Land	MOD	NEW	Land	MOD	NEW	Land	MOD	NEW	Land	MOD	NEW	Land	MOD	NEW
2002–2003	0.007	0.018	0.012	0.009	0.032	0.021	0.062	0.086	0.073	0.018	0.035	0.028	0.062	0.089	0.069	0.036	0.056	0.043
2003–2004	0.200	0.039	0.027	0.260	0.061	0.040	0.284	0.099	0.084	0.166	0.076	0.061	0.260	0.112	0.086	0.214	0.078	0.060
2004–2005	0.057	0.028	0.020	0.131	0.063	0.041	0.207	0.090	0.076	0.128	0.057	0.045	0.268	0.128	0.099	0.182	0.077	0.060
2005–2006	0.089	0.075	0.053	0.136	0.113	0.073	0.158	0.104	0.088	0.129	0.100	0.080	0.200	0.132	0.102	0.154	0.102	0.079
2006–2007	0.039	0.045	0.032	0.089	0.068	0.044	0.220	0.121	0.103	0.101	0.073	0.058	0.250	0.120	0.093	0.166	0.086	0.067
2007–2008	0.004	0.018	0.013	0.007	0.025	0.016	0.063	0.074	0.062	0.027	0.042	0.034	0.088	0.079	0.061	0.048	0.051	0.040
2008–2009	0.001	0.012	0.008	0.003	0.011	0.007	0.047	0.052	0.043	0.014	0.023	0.018	0.062	0.057	0.044	0.034	0.034	0.027
2009–2010	0.184	0.093	0.065	0.191	0.126	0.081	0.293	0.174	0.149	0.160	0.105	0.083	0.348	0.201	0.155	0.252	0.137	0.106
2010–2011	0.002	0.058	0.041	0.005	0.100	0.064	0.059	0.096	0.081	0.017	0.078	0.062	0.072	0.128	0.099	0.040	0.085	0.066
2011–2012	0.020	0.036	0.025	0.038	0.075	0.049	0.163	0.131	0.112	0.077	0.074	0.059	0.220	0.142	0.110	0.129	0.094	0.073
2012–2013	0.016	0.047	0.033	0.020	0.072	0.046	0.070	0.169	0.144	0.036	0.130	0.103	0.079	0.143	0.110	0.052	0.113	0.088

Table 3. Annual maximum average snow cover fraction (SCF) calculated for the three snow products, Landsat-mix (Land), MODIS-EURAC (MOD) and Corrected MODIS-EURAC (NEW) in each of the study regions and the whole study area.

	Region 1			Region 2			Region 3			Region 4			Region 5			Total		
	Land	MOD	NEW	Land	MOD	NEW	Land	MOD	NEW	Land	MOD	NEW	Land	MOD	NEW	Land	MOD	NEW
2002–2003	0.011	0.306	0.214	0.024	0.419	0.270	0.188	0.549	0.472	0.082	0.457	0.362	0.259	0.643	0.496	0.146	0.449	0.348
2003–2004	0.467	0.745	0.521	0.501	0.639	0.412	0.800	0.969	0.833	0.552	0.791	0.626	0.690	0.781	0.602	0.616	0.732	0.567
2004–2005	0.160	0.734	0.513	0.400	0.853	0.550	0.647	0.672	0.577	0.338	0.656	0.519	0.654	0.840	0.648	0.474	0.742	0.575
2005–2006	0.541	0.955	0.668	0.701	0.965	0.622	0.790	0.803	0.690	0.582	0.972	0.770	0.883	0.878	0.677	0.739	0.893	0.692
2006–2007	0.115	0.864	0.605	0.257	0.737	0.475	0.408	0.917	0.789	0.234	0.694	0.549	0.542	0.884	0.681	0.360	0.773	0.599
2007–2008	0.027	0.498	0.348	0.043	0.556	0.359	0.204	0.757	0.651	0.106	0.879	0.696	0.261	0.897	0.691	0.131	0.689	0.534
2008–2009	0.006	0.472	0.330	0.027	0.353	0.228	0.193	0.700	0.602	0.086	0.382	0.303	0.304	0.572	0.441	0.162	0.368	0.285
2009–2010	0.471	0.857	0.599	0.579	0.938	0.604	0.628	0.911	0.784	0.362	0.628	0.497	0.750	0.908	0.700	0.558	0.778	0.603
2010–2011	0.015	0.986	0.690	0.021	0.895	0.577	0.256	0.877	0.754	0.091	0.981	0.777	0.249	0.883	0.681	0.142	0.831	0.644
2011–2012	0.051	0.421	0.294	0.119	0.581	0.374	0.323	0.832	0.716	0.183	0.563	0.446	0.411	0.851	0.656	0.262	0.659	0.511
2012–2013	0.054	0.552	0.387	0.144	0.559	0.360	0.279	0.930	0.800	0.206	0.917	0.727	0.396	0.823	0.634	0.254	0.701	0.543

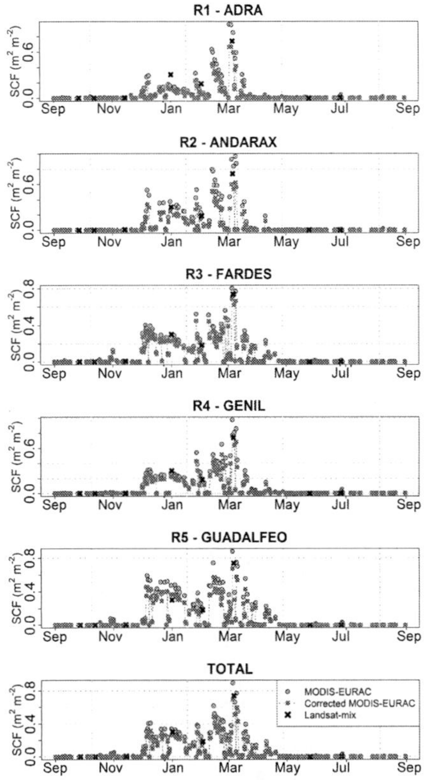

Figure 4. Example of reconstruction of MODIS-EURAC snow cover map for the year 2004–2005.

EURAC-MODIS product presented a general overestimation during the high snow covered period, which was partially solved with the correction with the linear model. Further ongoing work is exploring the apparent threshold in the large SCF values domain that can be observed in the graphs, together with the different behaviour of Region 5, the mostly-influenced by snow in the study area.

5 Conclusions

This work shows how the setting out of a simple approach provides a more accurate evolution of the average SCF values over this Mediterranean region, combining the advantages of two already existing products, the high spatial accuracy of Landsat-mix and the daily temporal resolution of MODIS-EURAC. The result is a daily time series on which different studies that require high resolution both of time and space can be based on. This work constitutes the first step in a more complex development of a data fusion algorithm that not only reproduces average behaviour but also snow distribution at grid/subgrid scales.

Data availability. The improved-MODIS snow products used in this study are available at http://sdi.eurac.edu/geonetwork/srv/eng/metadata.show?id=357955&currTab=advanced (Notarnicola et al., 2013a, b).

Author contributions. RP has been the main responsible of the calculation and writting; CM, LDG and CN have processed the improved-MODIS product; MJPP has contribute in the Landsat scene processing; and MJP has provided the initial idea of this study and help in the writting. All the authors have contributed in the review process.

Competing interests. The authors declare that they have no conflict of interest.

Special issue statement. This article is part of the special issue "Earth Observation for Integrated Water and Basin Management: New possibilities and challenges for adaptation to a changing environment". It is a result of The Remote Sens. & Hydrology Symposium, Cordoba, Spain, 8–10 May 2018.

Acknowledgements. This work was funded by the Spanish Ministry of Economy and Competitiveness – MINECO (Research Project CGL2014-58508R, "Global monitoring system for snow areas in Mediterranean regions: trends analysis and implications for water resource management in Sierra Nevada"). Moreover, this research was partially developed within the framework of the Panta Rhei Research Initiative of the International Association of Hydrological Science (IAHS) in the Working Group on Water and Energy Fluxes in a Changing Environment.

Edited by: Ana Andreu
Reviewed by: two anonymous referees

References

Andreadis, K. M. and Lettenmaier, D. P.: Assimilating remotely sensed snow observations into a macroscale hydrology model, Adv. Water Resour., 29, 872–886, https://doi.org/10.1016/j.advwatres.2005.08.004, 2006.

Barnett, T. P., Adam, J. C., and Lettenmaier, D. P.: Potential impacts of a warming climate on water availability in snow-dominated regions, Nature, 438, 303–309, 2005.

Dozier, J. and Painter, T. H.: Multispectral and Hyperspectral Remote Sens. of Alpine Snow Properties, Annu. Rev. Earth Pl. Sc., 32, 465–494, https://doi.org/10.1146/annurev.earth.32.101802.120404, 2004.

Giorgi, F.: Climate change hot-spots, Geophys. Res. Lett., 33, L08707, https://doi.org/10.1029/2006GL025734, 2006.

Hall, D. K. and Riggs, G. A.: Accuracy assessment of the MODIS snow products, Hydrol. Process., 21, 1534–1547, https://doi.org/10.1002/hyp.6715, 2007.

Hall, D. K., Riggs, G. A., Salomonson, V. V., DiGirolamo, N. E., and Bayr, K. J.: MODIS snow-cover products, Remote Sens. Environ., 83, 181–194, https://doi.org/10.1016/S0034-4257(02)00095-0, 2002.

Huang, J., Ji, M., Xie, Y., Wang, S., He, Y. and Ran, J.: Global semi-arid climate change over last 60 years, Clim. Dynam., 46, 1131–1150, https://doi.org/10.1007/s00382-015-2636-8, 2016.

Ménard, C. B., Essery, R., and Pomeroy, J.: Modelled sensitivity of the snow regime to topography, shrub fraction and shrub height, Hydrol. Earth Syst. Sci., 18, 2375–2392, https://doi.org/10.5194/hess-18-2375-2014, 2014.

Notarnicola, C., Duguay, M., Moelg, N., Schellenberger, T., Tetzlaff, A., Monsorno, R., Costa, A., Steurer, C., and Zebisch, M.: Snow Cover Maps from MODIS Images at 250 m Resolution, Part 1: Algorithm Description, Remote Sens., 5, 110–126, https://doi.org/10.3390/rs5010110, 2013a (data available at: http://sdi.eurac.edu/geonetwork/srv/eng/metadata.show?id=357955&currTab=advanced, last access: 11 September 2018).

Notarnicola, C., Duguay, M., Moelg, N., Schellenberger, T., Tetzlaff, A., Monsorno, R., Costa, A., Steurer, C. and Zebisch, M.: Snow Cover Maps from MODIS Images at 250 m Resolution, Part 2: Validation, Remote Sens., 5, 1568–1587, https://doi.org/10.3390/rs5041568, 2013b (data available at: http://sdi.eurac.edu/geonetwork/srv/eng/metadata.show?id=357955&currTab=advanced, last access: 11 September 2018).

Parajka, J. and Blöschl, G.: The value of MODIS snow cover data in validating and calibrating conceptual hydrologic models, J. Hydrol., 358, 240–258, https://doi.org/10.1016/j.jhydrol.2008.06.006, 2008.

Pérez-Palazón, M. J., Pimentel, R., Herrero, J., Aguilar, C., Perales, J. M., and Polo, M. J.: Extreme values of snow-related variables in Mediterranean regions: trends and long-term forecasting in Sierra Nevada (Spain), Proc. IAHS, 369, 157–162, https://doi.org/10.5194/piahs-369-157-2015, 2015.

Pimentel, R., Herrero, J., Zeng, Y., Su, Z., and Polo, M. J.: Study of Snow Dynamics at Subgrid Scale in Semiarid Environments Combining Terrestrial Photography and Data Assimilation Techniques, J. Hydrometeor., 16, 563–578, https://doi.org/10.1175/JHM-D-14-0046.1, 2015.

Pimentel, R., Herrero, J., and Polo, M. J.: Quantifying Snow Cover Distribution in Semiarid Regions Combining Satellite and Terrestrial Imagery, Remote Sens., 9, 995, https://doi.org/10.3390/rs9100995, 2017a.

Pimentel, R., Herrero, J., and Polo, M. J.: Subgrid parameterization of snow distribution at a Mediterranean site using terrestrial photography, Hydrol. Earth Syst. Sci., 21, 805–820, https://doi.org/10.5194/hess-21-805-2017, 2017b.

Roy, D. P., Wulder, M. A., Loveland, T. R., C.e., W., Allen, R. G., Anderson, M. C., Helder, D., Irons, J. R., Johnson, D. M., Kennedy, R., Scambos, T. A., Schaaf, C. B., Schott, J. R., Sheng, Y., Vermote, E. F., Belward, A. S., Bindschadler, R., Cohen, W. B., Gao, F., Hipple, J. D., Hostert, P., Huntington, J., Justice, C. O., Kilic, A., Kovalskyy, V., Lee, Z. P., Lymburner, L., Masek, J. G., McCorkel, J., Shuai, Y., Trezza, R., Vogelmann, J., Wynne, R. H., and Zhu, Z.: Landsat-8: Science and product vision for terrestrial global change research, Remote Sens. Environ., 145, 154–172, https://doi.org/10.1016/j.rse.2014.02.001, 2014.

Proc. IAHS, 380, 73–79, 2018
https://doi.org/10.5194/piahs-380-73-2018

Open Access
PIAHS

Earth Observation for Integrated Water and Basin Management: Challenges for adaptation to a changing environment

Use of Sentinel 2 – MSI for water quality monitoring at Alqueva reservoir, Portugal

Miguel Potes[1], **Gonçalo Rodrigues**[1], **Alexandra Marchã Penha**[1], **Maria Helena Novais**[1], **Maria João Costa**[1,2], **Rui Salgado**[1,2], **and Maria Manuela Morais**[1,3]

[1]Institute of Earth Sciences – ICT, IIFA, University of Évora, Évora, 7000-671, Portugal
[2]Department of Physics, ECT, University of Évora, Évora, 7000-671, Portugal
[3]Department of Biology, ECT, University of Évora, Évora, 7000-671, Portugal

Correspondence: Miguel Potes (mpotes@uevora.pt)

Received: 18 April 2018 – Revised: 10 July 2018 – Accepted: 27 July 2018 – Published: 18 December 2018

Abstract. Alqueva reservoir located in southeast of Portugal has a surface area of 250 km^2 and total capacity of 4150 hm^3. Since 2006 the water quality of this reservoir is explored by the authors using remote sensing techniques. First using MERIS multi-spectral radiometer on-board of ENVISAT-1 and presently with MSI multi-spectral radiometer on-board SENTINEL-2. The existence of two satellites (A and B) equipped with MSI enable the area to be revisited, under the same viewing conditions, every 2–3 days. Since 2017 the multidisciplinary project ALOP (ALentejo Observation and Prediction systems) expands the team knowledge about the physical and bio-chemical properties of the reservoir. This project includes an integrated field campaign at different experimental sites in the reservoir and its shores, at least until September 2018. Previous algorithms developed by the team for MERIS are tested with the new MSI instrument for water turbidity, chlorophyll *a* concentration and density of cyanobacteria. Results from micro-algae bloom occurred in late summer/early autumn 2017 on the reservoir are presented, showing the capabilities of MSI sensor for detection and high resolution mapping over the reservoir. The results are compared with in situ sampling and laboratorial analysis of chlorophyll *a* associated with the bloom.

1 Introduction

The water quality monitoring of inland reservoirs is essential to ensure that it remains within acceptable boundaries or otherwise, to trigger actions that may revert the water quality degradation. The set-up of early warning systems is important for reservoirs, which may be affected by pollution events, alerting the authorities to water quality degradation episodes. The Alqueva reservoir, located in southwestern Iberian Peninsula (south of Portugal), has a total capacity of 4.150 hm^3 and a surface area of 250 km^2, constituting the largest reservoir in the Iberian Peninsula. It is a multipurpose structure used for water supply, irrigation, hydroelectric power generation and recreation, thus water quality management is critical. Since 2017 the multidisciplinary project ALOP (ALentejo Observation and Prediction systems) expands the knowledge about the physical and biochemical properties of Alqueva reservoir. This project includes an integrated field campaign at different experimental sites in the reservoir and its shores, at least until September 2018. It aims to develop a multi-functional activity in the field of atmosphere-water-ecosystem interaction, which embraces observation, prediction and risk alert. It intends to develop tools of observation, forecasting and alert in the domains of meteorology and water (quantity and quality), at regional scale. Satellite remote sensing constitutes a useful tool to complement in situ data sampling and laboratorial analysis of water quality parameters, which is laborious and expensive and thus spatially and temporally limited. Unless a water body is adequately instrumented with in situ sensors, remote-sensing is the only suitable method to monitor the quality of remote and large inland waters (WMO, 2013) and can undoubtedly contribute to early warning systems. Several satellite remote sensing studies have been proposed using a variety of sensors over a number of worldwide reservoirs (Gholizadeh et al., 2016). Remote sensing of water bodies rely on the varying colour of natural waters that correspond to different spectral reflectances, assuming that these variations depend on the water constituents. Water remote sensing was used to monitor ocean colour already since the 1960s, and in the last three decades there has been a growing interest to apply these remote sensing techniques also to inland water quality. For this purpose, remote sensing measurements, mainly using visible and near-infrared wavelengths, are used to develop bio-optical models that aim to relate radiometric (optical) and biological quantities. Inland waters are optically complex due to the presence of several constituents that may interact, creating uncertainties in the remote sensing retrievals (Toming et al., 2016; Ogashawara et al., 2017). Thus there is a need to study inland waters at a local and regional scale and to quantify the performance of remote sensing methods to monitor inland water quality. The first remote sensing studies over Portuguese reservoirs were carried out by the authors, proposing semi-empirical bio-optical models to estimate concentrations of chlorophyll *a* and cyanobacteria (Potes et al., 2011), as well as turbidity (Potes et al., 2012) over Alqueva from MERIS sensor onboard ENVISAT satellite.

The launch of ESA Sentinel-2 (MSI) mission (S-2A launched in June 2015 and S-2B launched in March 2017), carrying as single payload the Multi-Spectral Instrument (MSI), brought a great opportunity to study inland reservoirs. It presents a systematic global coverage of two to three days at mid-latitudes, which supports monitoring of changes in reservoirs, with relatively high spatial resolution. MERIS sensor, on board ENVISAT, presented maximum spatial resolution of 300 m, whereas MSI presents a spatial resolution of 10, 20 or 60 m, depending on the spectral band. Potes et al. (2011, 2012) proposed a method to monitor chlorophyll *a*, cyanobacteria and turbidity over inland reservoirs using MERIS data. The objective of this work is to show the potential of MSI used with the set of algorithms developed for MERIS, to study the water quality of inland waters. Specifically, a micro-algae bloom that occurred in early autumn 2017 is analysed and its evolution related to the influence of a hurricane progressing over the Eastern North Atlantic region.

Section 2 presents the data used and methodology. Results are presented and discussed in Sect. 3 and finally Sect. 4 summarizes the main conclusions.

2 Data and methods

Data from Sentinel 2 is used in this work, namely level 2A from MSI instrument, which corresponds to images with atmospheric correction, providing information of surface reflectance in 12 spectral channels from 443 to 2190 nm (central wavelength). The product Maximum Chlorophyll Index (MCI) was also extracted from the level 2 images which is an indicator of the amount of chlorophyll present in water mass (Gower et al., 2008) and thus a useful tool in the monitoring of algae blooms of inland waters. The bands selected for MCI index were the 665, 705 and 740 nm.

Table 1. MERIS bands wavelengths and equivalent MSI bands used in the algorithms.

MERIS		MSI	
Wavelength (nm)	Bandwidth (nm)	Wavelength (nm)	Bandwidth (nm)
412.5	10	443	20
442.5	10	443	20
490	10	490	65
560	10	560	35
620	10	665	30

The algorithms from Potes et al. (2011, 2012) developed for MERIS are:

$$\text{Chl}\ a = 4.23\left(\frac{560\,\text{nm}}{442.5\,\text{nm}}\right)^{3.94}\ \left[\text{mg}\,\text{m}^{-3}\right] \quad (1)$$

$$\text{Cya} = 115\,530.31\left(\frac{560\,\text{nm}\cdot 620\,\text{nm}}{490\,\text{nm}}\right)^{2.38}\ \left[10^{3}\,\text{cells}\,\text{mL}^{-1}\right] \quad (2)$$

$$\text{Turb} = 8.93\left(\frac{560\,\text{nm}}{412.5\,\text{nm}}\right) - 6.39\ [\text{NTU}] \quad (3)$$

Table 1 presents the information about MERIS bands used in the algorithms to estimate the concentration of chlorophyll *a* (Chl *a* in Eq. 1), density of cyanobacteria (Cya in Eq. 2) and water turbidity (Turb in Eq. 3). In the same table the equivalent MSI wavelengths applied in this work is presented.

The algorithms represented by Eqs. (1) to (3) were applied now to the same water body as in Potes et al. (2011, 2012). The chlorophyll *a* concentration was also obtained on a bimensal basis from in situ sampling and laboratory analysis in the framework of ALOP project, in order to validate the results. The method used was the molecular absorption spectroscopy and the equations developed by Lorenzen (1967).

3 Study site and case study

Alqueva reservoir project was concluded in 2002 and the reservoir reached the capacity of 80 % in March 2004. Figure 1 shows the surface area of the reservoir at its full capacity as well as the sites considered in this work.

The algorithms developed by Potes et al. (2011, 2012) for MERIS have been applied to MSI imagery in Alqueva reservoir. ALOP field campaign is ongoing since January 2017 and a case study of October 2017 was chosen, because this was a particular month in Portugal with the influence of Ophelia hurricane amidst a very dry period. Ophelia started as a tropical storm but it reached category 3 hurricane in Saffir–Simpson hurricane wind scale, South of Azores islands being the strongest hurricane ever recorded so far East

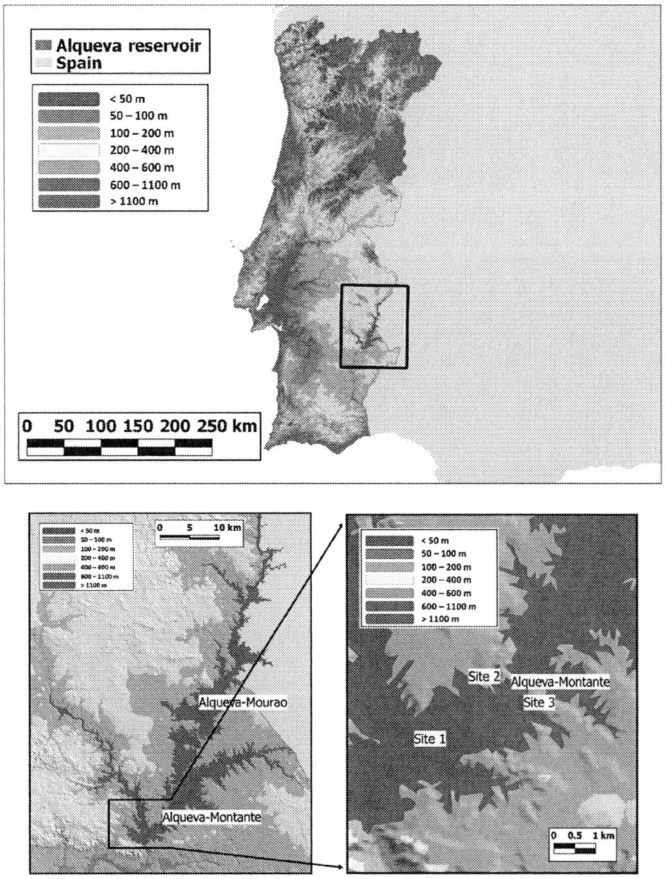

Figure 1. Map of Alqueva reservoir located in Southeast Portugal. The sites used in this work are also represented.

in the Atlantic (Fig. 2). In its trajectory towards Ireland it started to lose strength as is entered in cold waters being very close to Iberian Peninsula, about 360 km from Cape Finisterre (Spain).

This hurricane brought a meteorological breakup window in the unprecedented drought scenario that the country was facing by October (which is normally the beginning of the rainy season). In particular, this event led to the increase of wind speed and relative humidity and a decrease in the air temperature. Precipitation was concentrated in two days (17 and 18 October) with an accumulated value of 32.0 mm. Figure 3 shows the evolution of air temperature for October 2017 recorded in CidAlmeida meteorological station installed in the area of Alqueva (ALOP project) where the influence of the hurricane can be detected between 16 and 22 October.

Figure 4 presents a wind rose from the same inland station for the period 15 to 22 October, where the stronger winds from the South and West quadrants are clearly visible, as a consequence of hurricane Ophelia passing by Portugal.

According to this scenario, three Sentinel-2 images from October 2017 were selected on clear sky days (12, 22 and 29) to use the MSI instrument in Alqueva reservoir. Table 2 shows acquisitions date and time for both satellite data and in situ water collection.

4 Results

On 12 October the reservoir was under a micro-algae bloom especially on the northern part as can be seen in Fig. 5 (row a) in terms of concentration of chlorophyll *a*, density of cyanobacteria and water turbidity. Values greater than 50 mg m^{-3} of chlorophyll *a* were estimated for the northern part, as well as greater than 75 000 cells mL^{-1} cyanobacteria with turbidity values reaching 30 NTU.

On 22 October (Fig. 5 – row b), six days after, all the three parameters decreased sharply in the northern part probably due to very different weather conditions in the six days under the influence of the hurricane Ophelia passage along

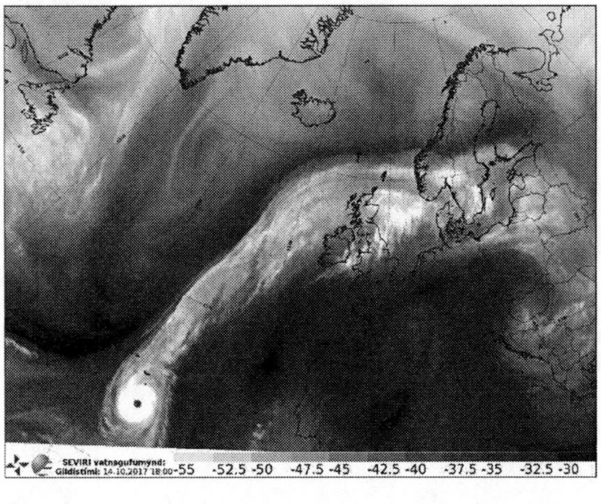

Figure 2. Spinning Enhanced Visible and Infrared Imager (SEVIRI) image of hurricane Ophelia located South of Azores islands on 14 October 2017, at 18:00 UTC (https://irishweatheronline.wordpress.com/, last access: 15 April 2018).

Table 2. MSI acquisition dates and in situ date for water collection.

Date	Time (UTC)	Type
12 Oct 2017	11:21	Satellite
22 Oct 2017	11:21	Satellite
27 Oct 2017	11:00–14:00	Laboratory
29 Oct 2017	11:12	Satellite

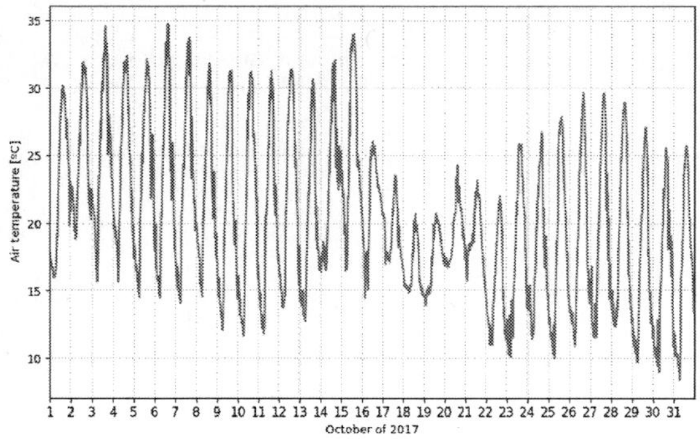

Figure 3. One-minute averages of air temperature recorded in CidAlmeida meteorological station located inland nearby Alqueva-Montante site (Fig. 1) for October 2017.

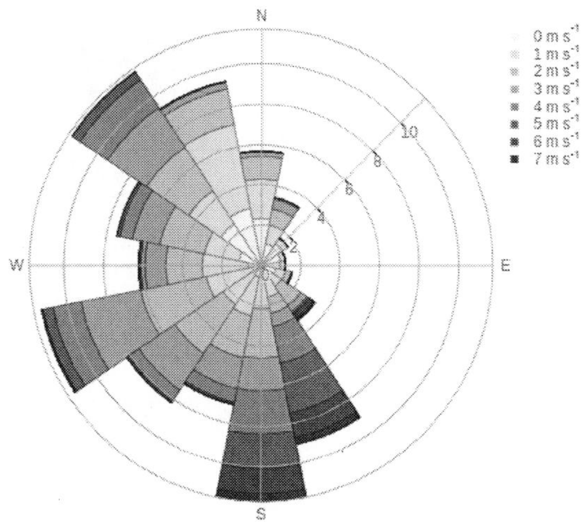

Figure 4. Wind rose recorded in a meteorological station located inland nearby Alqueva-Montante site (Fig. 1) for the period 15 to 22 October 2017.

the Portuguese coast. It is well known that lower temperatures inhibit the phytoplankton growth (Wu et al., 2015) the same occurs with increasing wind speed and precipitation, which leads to increasing in water mixing (Fleming-Lehtinen and Laamanen, 2012). For these reasons a regress of the bloom due to the changes in meteorological conditions was expected.

In the central part of the reservoir the values are above the eutrophic threshold for chlorophyll *a* presented by Bukata et al. (1995), of $6\,\mathrm{mg\,m^{-3}}$. In addition, most of the branches of reservoir present values are above the ecological potential and eutrophic threshold for chlorophyll *a* for reservoirs in the South of Portugal, which according with the Water Frame Directive is 9.5 and $8\,\mathrm{mg\,m^{-3}}$, respectively (INAG, 2009). Nine days after 22 October, on the images of 29 October presented in Fig. 5 (row c) the reservoir is clear from the presence of the micro-algae bloom. All the three parameters present low values indicating good water quality on that day. Nevertheless, some thinner branches of the reservoir still present some occasional high values. A summary table with minimum, maximum and mean values is presented for the three parameters in the three days (Table 3).

Figure 6 presents a scatter plot between the MCI and concentration of chlorophyll *a* obtained from satellite. All pixels from 22 October are used in the plot, in a total of 28 251 points. These two parameters are well correlated (correlation coefficient of 0.70) with a RMSE (Root Mean Square Error) of $21.5\,\mathrm{mg\,m^{-3}}$. Future work foresees the use of longer data series to quantify MCI as chlorophyll concentration.

Satellite derived results were compared with data obtained from laboratory analysis. Unfortunately, there are no water laboratory analyses for the same days as MSI acquired images, rather five days before (22 October) and two days after (29 October). From the analysis of MSI retrieved data, it was concluded that the water quality has improved from 22 to 29 October and thus it is expected that the water analysis should be more or less similar or in-between the values of MSI. This is shown in Fig. 7 where the chlorophyll *a*

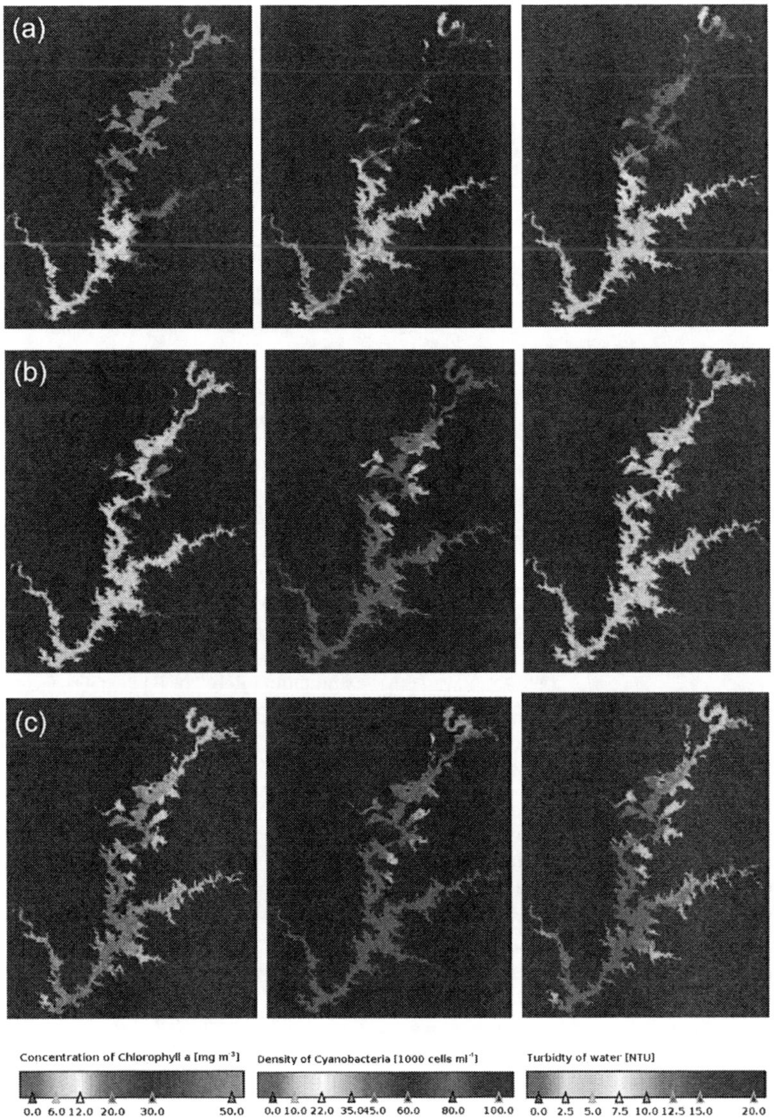

Figure 5. Concentration of chlorophyll *a*, density of cyanobacteria and water turbidity estimated from the Eqs. (1) to (3) for 12 October (row **a**), 22 October (row **b**) and 29 October (row **c**).

concentration estimated from MSI (22 and 29 October) and obtained in laboratory (27 October) for five sites (Fig. 1) is plotted. The results are generally in agreement. From 22 to 29 October, MSI reports a decrease in chlorophyll *a* concentration and the laboratory analysis (middle day) confirms this decrease presenting a middle value in all sites, except for Alqueva-Montante and Site 1, which presents a very similar value nevertheless slightly greater than the MSI value for 22 October. In these two sites the MSI values, nearby the in situ measurement place, were more heterogeneous than in the other sites probably due to their central location where the water is under more currents than in the other sites, in this part of the year. The results also suggest that the decrease of chlorophyll *a* was not linear in time, but more pronounced near the end of the five day period.

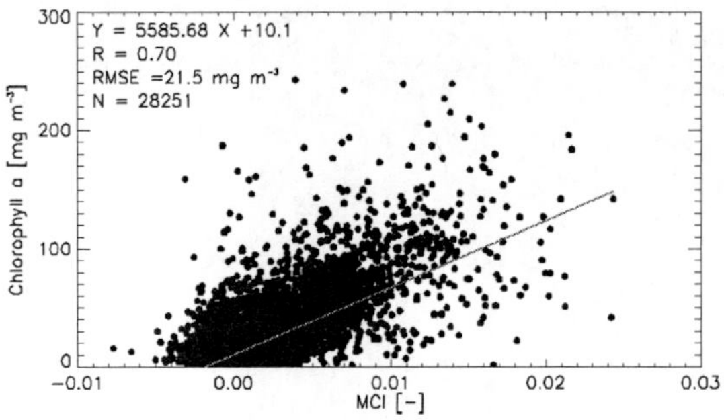

Figure 6. Scatter plot between MCI and concentration of chlorophyll *a* estimated by Eq. (1) for day 22 October all pixels.

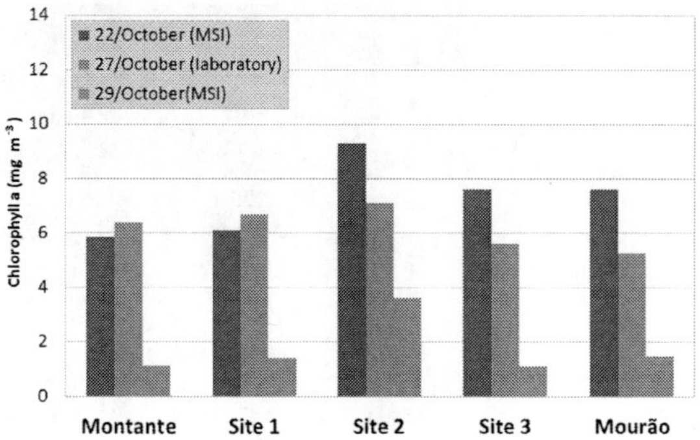

Figure 7. Bar plot of concentration of chlorophyll *a* estimated by Eq. (1) for 22 and 29 October and obtained in laboratory for 27 October for 5 sites shown on Fig. 1.

Table 3. Range and mean values of the maps presented in Fig. 5.

Date		Chlorophyll *a* [$mg\,m^{-3}$]	Cyanobacteria [10^3 cells mL^{-1}]	Turbidity [NTU]
12 Oct	Range	[2.6; 600.1]	[0.1; 228.1]	[1.5; 25.0]
	Mean	46.1	33.3	8.6
22 Oct	Range	[0.5; 263.7]	[0.2; 283.0]	[0.1; 19.1]
	Mean	14.1	4.0	4.5
29 Oct	Range	[0.1; 140.1]	[0.1; 179.7]	[0.3; 15.3]
	Mean	4.6	3.8	1.2

5 Conclusions

In this work the potential of Sentinel-2 MSI instrument for monitoring the water quality of inland waters is shown. In the case study reported, different conditions of Alqueva reservoir are analysed for one month. Systems like Alqueva reservoir are highly dynamic and respond rapidly to changes in atmospheric conditions, thus is very importance to have these kinds of tools for monitoring the water surface as a whole. In this work, a set of algorithms developed for another space borne spectroradiometer (MERIS) were applied with good results. The team is now working to tune the algorithms in order to make use of the full potential of MSI instrument. Specific algorithms were already developed successfully by Toming et al. (2016) using the same MSI instrument over Estonian lakes, regarding chlorophyll *a* concentration as well as colored dissolved organic matter (CDOM) and dissolved organic carbon (DOC).

Data availability. Data used in this work is available on request to the first author.

Author contributions. MP conceptualized and led the team on this research. He also wrote great part of the manuscript and carried out the corrections during the write-up. GR gathered, processed and prepared the spatial data and maps used for this article. AMP and MHN were responsible for the in situ and laboratory data analysis used in the work. MJC participated in the data processing and discussion and wrote the first part of the manuscript. RS and MMM participated in the discussion of results and reviewed the manuscript for necessary corrections.

Competing interests. The authors declare that they have no conflict of interest.

Special issue statement. This article is part of the special issue "Earth Observation for Integrated Water and Basin Management: New possibilities and challenges for adaptation to a changing environment". It is a result of The Remote Sensing & Hydrology Symposium, Cordoba, Spain, 8–10 May 2018.

Acknowledgements. The work was funded by the ALOP project (ALT20-03-0145-FEDER-000004) and also through the European Union through the European Regional Development Fund, included in the COMPETE 2020 (Operational Program Competitiveness and Internationalization) through the ICT project (UID/GEO/04683/2013) with the reference POCI-01-0145-FEDER 007690.

Edited by: María José Polo
Reviewed by: two anonymous referees

References

Bukata, R. P., Jerome, J. H., Kondratyev, K. Y., and Pozdnyakov, D. V.: Optical Properties and Remote Sensing of Inland and Coastal Waters, 135–250, CRS Press, Boca Raton, Florida 33431, USA, 1995.

Fleming-Lehtinen, V. and Laamanen, M.: Long-term changes in Secchi depth and the role of phytoplankton in explaning light attenuation in the Baltic Sea, Estuar. Coast. Shelf S., 102–103, 1–10, 2012.

Gholizadeh, M. H., Melesse, A. M., and Reddi, L.: A Comprehensive Review on Water Quality Parameters Estimation Using Remote Sensing Techniques, Sensors, 16, 1298, https://doi.org/10.3390/s16081298, 2016.

Gower, J. F., King, R. S., and Goncalves, P.: Global monitoring of plankton blooms using MERIS MCI, Int. J. Remote Sens., 29, 6209–6216, https://doi.org/10.1080/01431160802178110, 2008.

INAG (Instituto da Água Instituto Público): Critérios para a Classificação do Estado das Massas de Água Superficiais – Rios e Albufeiras, Ministério do Ambiente, do Ordenamento do Território e do Desenvolvimento Regional, Lisboa, Portugal, 2009.

Lorenzen, C. J.: Determination of chlorophyll and phaeopigments: Spectrophotometric equations, Limnol. Oceanogr., 12, 348–356, 1967.

Ogashawara, I., Mishra, D. R., and Gitelson, A. A.: Chapter 1 – Remote Sensing of Inland Waters: Background and Current State-of-the-Art, in: Bio-optical Modeling and Remote Sensing of Inland Waters, edited by: Mishra, D. R., Ogashawara, I., and Gitelson, A. A., Elsevier, 1–24, https://doi.org/10.1016/B978-0-12-804644-9.00001-X, 2017.

Potes, M., Costa, M. J., Silva, J. C. B., Silva, A. M., and Morais, M.: Remote sensing of water quality parameters over Alqueva reservoir in the south of Portugal, Int. J. Remote Sens., 32, 3373–3388, https://doi.org/10.1080/01431161003747513, 2011.

Potes, M., Costa, M. J., and Salgado, R.: Satellite remote sensing of water turbidity in Alqueva reservoir and implications on lake modelling, Hydrol. Earth Syst. Sci., 16, 1623–1633, https://doi.org/10.5194/hess-16-1623-2012, 2012.

Toming, K., Kutser, T., Laas, A., Sepp, M., Paavel, B., and Nõges, T.: First Experiences in Mapping Lake Qater Quality Parameters with Sentinel-2 MSI Imagery, Remote Sens., 8, 640, https://doi.org/10.3390/rs8080640, 2016.

WMO: Planning of water-quality monitoring systems, WMO-No. 1113, Geneva, Switzerland, 2013.

Wu, Z., Zhang, Y., Zhou, Y., Liu, M., Shi, K., Yu, Z., and Lin, Y.-P.: Seasonal-Spatial Distribution and Long-Term Variation of Transparency in Xin'anjiang Reservoir: Implications for Reservoir Management, Int. J. Environ. Res. Pu., 12, 9492–9507, 2015.

Proc. IAHS, 380, 81–85, 2018
https://doi.org/10.5194/piahs-380-81-2018

Open Access PIAHS

Earth Observation for Integrated Water and Basin Management: Challenges for adaptation to a changing environment

Image acquisition effects on Unmanned Air Vehicle snow depth retrievals

Ahmet Emre Tekeli and Senayi Dönmez

Civil Engineering Department, Çankırı Karatekin University, Çankırı, 18100, Turkey

Correspondence: Ahmet Emre Tekeli (ahmetemretekeli@karatekin.edu.tr)

Received: 16 April 2018 – Revised: 20 September 2018 – Accepted: 24 September 2018 – Published: 18 December 2018

Abstract. Advancements in technology have facilitated new opportunities in aerial photogrammetry; one of these is the use of unmanned aerial vehicles (UAVs) to estimate snow depth (SD). Here, a multi-rotor type UAV is used for SD retrievals over an area of 172 000 m^2. Photos with 80 % forward and 60 % side overlaps were taken by UAV on two different (snow-covered and snow-free) days. SD estimations were obtained from the difference between 3-D stereo digital surface models (DSMs) produced for both days. Manual SD measurements were performed on the ground concurrent with UAV flights. The current study is unique in that the SD retrievals were derived using two different image acquisition modes. In the first, images were taken as UAV was continuously flying and in the second UAV had small stops and kept its position in air fixed as the photos were taken. Root mean square error of UAV derived SDs is calculated as 2.43 cm in continuous and 1.79 cm in fixed acquisitions. The results support the hypothesis, based on theoretical considerations, that fixed-position image acquisitions using multi-rotor platforms should enable more accurate SD estimates. It is further seen that, as SDs increased, the errors in SD calculations are reduced.

1 Introduction

Accurate estimation of water potential within the basin is important for optimum management of water resources. To this end, timely and accurate measurements of the rainfall and snowfall, which are major fresh water inputs into the basin, are needed. Runoff due to snowmelt is key to meeting the demands for freshwater in many regions (Barnett et al., 2005).

Reliable estimation of snow depth (SD) and snow water equivalent (SWE) which are indicators of water potential of the basin are very important for hydrological modelling, flood forecasting, avalanche mitigation and disaster management (Vander Jagt et al., 2015). SD and SWE measurements have been performed since 1960s using the snow courses. However, generally bi-weekly performed snow courses are not temporally dense enough to monitor changes in snow depth. Snow pillows and depth sensors improve the temporal resolutions of SD and SWE data obtained from the field. Moreover, they provide the most reliable information about SD and SWE. However, they provide point values and can only explain 30 % of the observed spatial variability in SD (Erxleben et al., 2002). Moreover, getting data from mountainous regions, where the main snowfall occurs, is limited due to safety and logistics.

Snow covered areas (SCA), by changing from 46.5×10^6 km in January to 3.8×10^6 km in August for Northern Hemisphere (Robinson et al., 1993), show the highest variation on Earth's surface after sea ice (Papa et al., 2002). Remote sensing (RS) provided new opportunities in globally monitoring SCA showing such large variation. Furthermore, RS facilitates the collection of temporally and spatially distributed SCA information while minimizing the risks associated with data acquisition with in-situ methods in high-risk areas (Vander Jagt et al., 2015).

Although methods for SD estimations using aerial photogrammetry have been available since the 1960s, the utility of these methods has been limited due to the high cost and limited accuracy of the SD estimates (Vander Jagt et al., 2015; Nolan et al., 2015). However, due to recent advancements in computer technology including the performance of photogrammetric software, along with improvement in cameras, Global Positioning System (GPS), and Inertial Measurement Units (IMU), there has been a renewed interest in the use of aerial photogrammetry for monitoring snow

Figure 1. Study area within Uluyazı Campus.

Figure 2. UAV used in the study.

depth. Mounting such equipment onto Unmanned Air Vehicles (UAV) has led to reduced aerial photogrammetry costs which eventually increased the use of UAVs in the field of geosciences (Westoby et al., 2012; Colomina and Molina, 2014). UAVs provided high temporal and spatial resolutions with their rapid image acquisition and low altitude flying capabilities. Moreover, as UAVs became portable, they have been used in cadastral, archeological and vegetation studies (Manyoky et al., 2011; Rinaudo et al., 2012; Zarco-Tejada et al., 2012) and started to take place in many applications of classical RS and aerial photogrammetry.

Investigations into the utility of using UAVs to estimate SD are nascent and prior studies have focused on the use of in-flight imagery from fixed-wing or multi-rotor UAVs (Bühler et al., 2016; Harder et al., 2016; De Michele et al., 2016). In this study, the effects of image acquisition method (UAV in-flight and UAV at fixed position) on UAV based snow depth accuracies are investigated. To the current knowledge of the authors such a comparison is not available in literature yet.

2 Study Area

The study area is located in Uluyazı Campus of Çankırı Karatekin University being in 3 km north-east of Çankırı province (40°37′ N, 33°36′ E). Figure 1 shows the study area (172 000 m^2) in red polygon roughly including Engineering and Science Faculties and the open space to the south. It comes fourth, in terms of areal coverage, after Harder et al. (2016), 320 000 m^2, De Michele et al. (2016), 300 000 m^2, Bühler et al. (2016), 275 000 m^2.

3 Methodology

SD estimations using UAV are based on the difference of Digital Surface Models (DSMs) obtained on snow covered and snow-free days (Vander Jagt et al., 2015; Bühler et al., 2016; De Michele et al., 2016; Avanzi et al., 2017). Although some researchers used fixed wing UAVs (De Michele et al., 2016; Harder et al., 2016), a multi-rotor UAV, also known as quadcopter or multicopter (Fig. 2), namely DJI Phantom, is used here because of its greater stability (Vander Jagt et al., 2015; Bühler et al., 2016; Avanzi et al., 2017).

To increase the accuracy of the derived DSMs and orthophotos, ground control points (GCPs) distributed as homogeneously as possible over the study area were used following the approach of Vander Jagt et al. (2015), De Michele et al. (2016) and Harder et al. (2016). The location (x, y, z) of GCPs were collected using Leica Viva Global Navigation Satellite Systems (GNSS) receivers.

The flight dates, 3 February and 24 March 2017 for the snow-covered and snow-free cases, respectively, were chosen based on the predicted total cloud cover from the ECMWF numerical weather prediction model. Flight path was arranged to enable 80 % forward and 60 % side overlap and uploaded to UAV before the take-off. Other than take-off and landing phases all flight was performed autonomously. Figure 3a and b shows photos from snow-covered and snow-free conditions of the yellow region presented in Fig. 1.

4 Results and Discussions

Before SD calculations, the geolocation (x, y, z) accuracies of DSMs were assessed by comparing with GCP measurements performed onsite. They were found to be accurate to within ±2.5 cm; this is similar to the uncertainty of 3 cm reported by Nolan et al. (2015). Then, orthophotos and DSMs were obtained for snow and snow free days (Fig. 4a and b).

Using the orthophotos and point clouds, DSMs (Fig. 5a and b) were derived. The difference between two DSMs produced SD values. The differences between UAV derived SDs and manual measurements are presented in Fig. 6 for two different image acquisitions (in-flight/at fixed point) on snowy day. Figure 6 shows that the bias in SD calculations decreases as SD values increase. This finding is similar to De Michele et al. (2016) and Avanzi et al. (2017). Although SD was determined for all locations using the imagery collected for UAV in-flight, it was not possible to determine SD in all cases

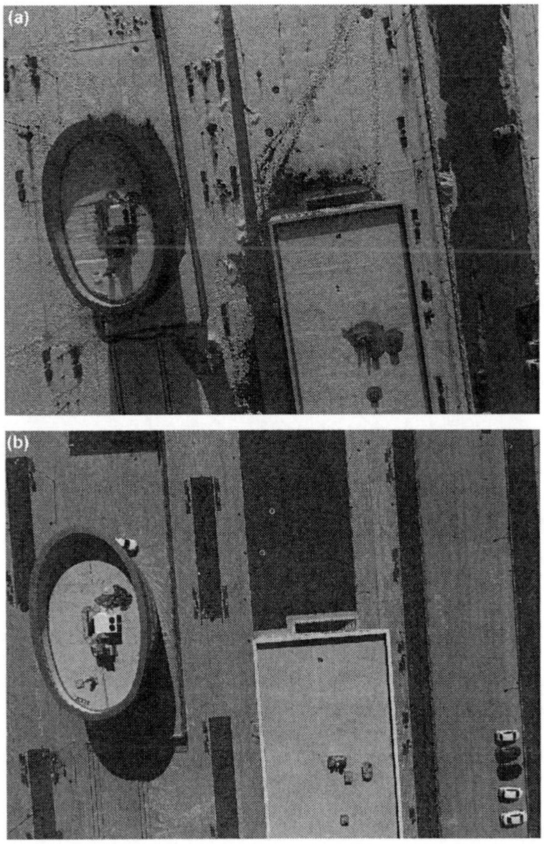

Figure 3. Image for 3 February (a) and 24 March 2017 (b).

when using imagery collected for UAV in fixed-positions due to the insufficient lighting.

Figure 7 shows the bias comparisons for the updated case based on the light conditions.

Scattered diagrams of calculated and measured SDs are presented in Fig. 8 for both in-flight/at fixed point cases.

Figures 7 and 8 indicate that the calculated SDs by fixed point image acquisitions gave smaller biases and higher correlations.

5 Conclusions

In this study, SDs derived from UAV images obtained by using two different image acquisition modes were compared. In the first case, the images were acquired while the UAV was in continuous flight mode and in the second case; they were collected while the UAV was kept in a fixed position. The UAV derived SDs were also compared with manual SD measurements performed concurrent with UAV flights over an area of 172 000 m^2 on the Uluyazı Campus of Çankırı Karatekin University. The locational accuracies DSMs were computed using ground control points (GCPs) as ±2.5 cm. The Root Mean Square Error (RMSE) of SD measurements were 2.43 and 1.79 cm, respectively, for in-flight and at fixed-position image acquisitions. The smaller RMSE agrees with the theoretical expectation. Also, the coefficient of determination (R^2) was higher for fixed position image acquisitions. Although the methodology used is the same in all cases, the accuracy of the SD estimates increases as the depth increased from 10.5 to 225 cm.

The same effect of lighting conditions on the accuracy of SD estimates described by Harder et al. (2016) were also observed here. Specifically, insufficient light reduced image contrast which eventually affected the accuracy and utility of SD estimates. This effect was particularly pronounced

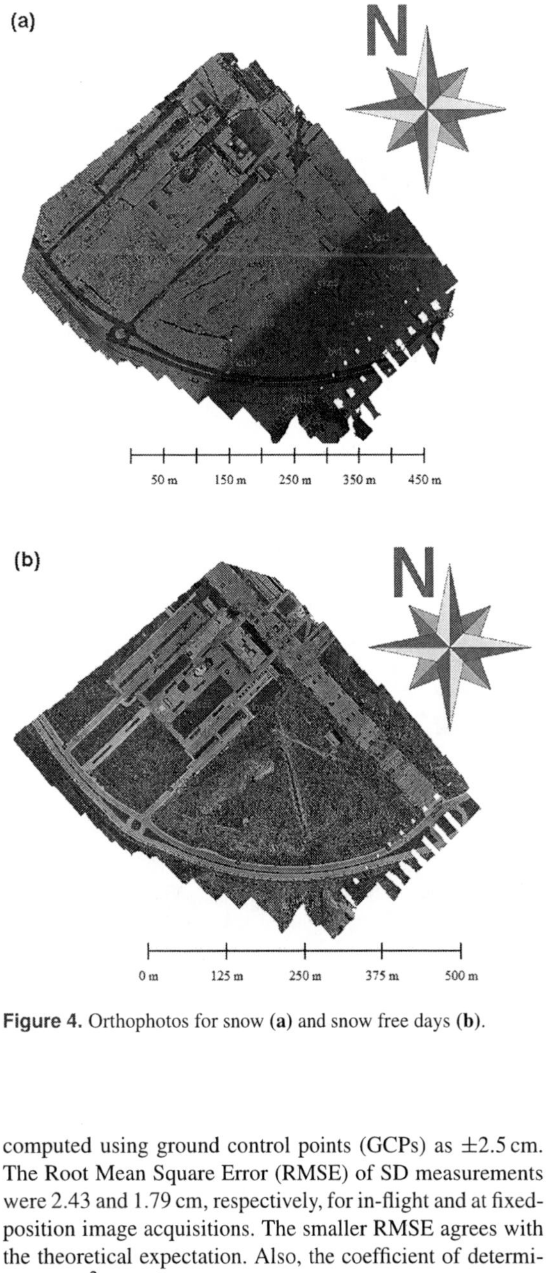

Figure 4. Orthophotos for snow (a) and snow free days (b).

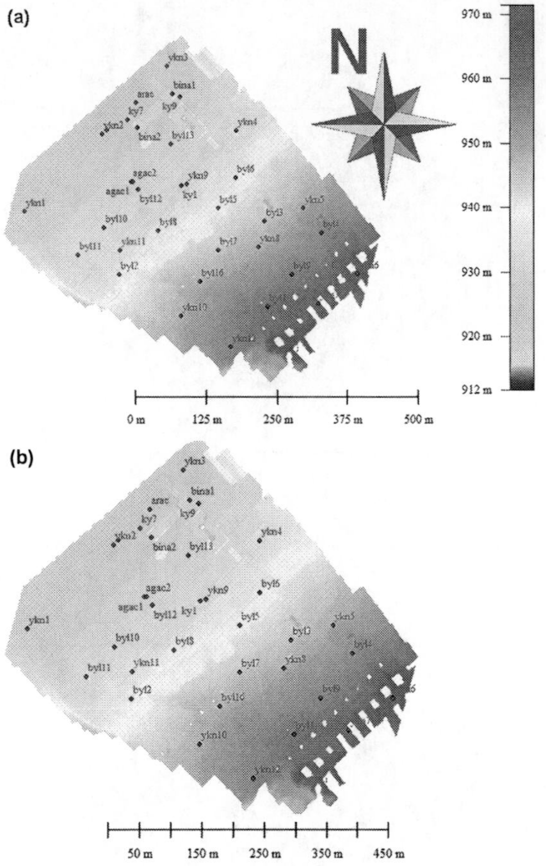

Figure 5. DSMs for snow-covered (**a**) and snow-free days (**b**).

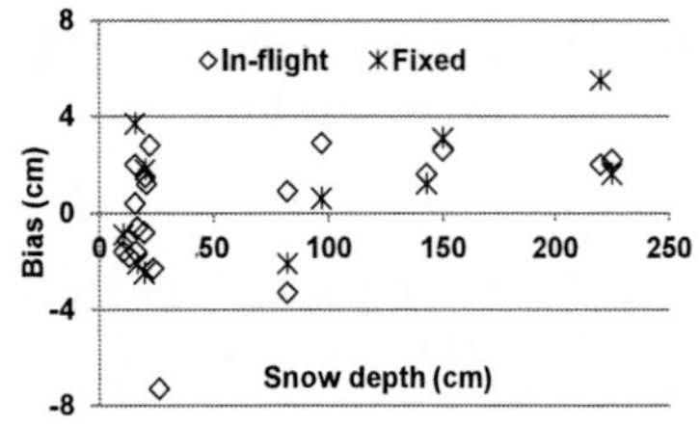

Figure 6. SD bias values for in-flight/at fixed point image acquisitions.

when the approach was applied to imagery collected while the UAV was in fixed-position image acquisition mode.

This work should be followed up by additional studies using near infrared imagery collected via UAV to estimate SD, as also discussed by Bühler et al. (2016).

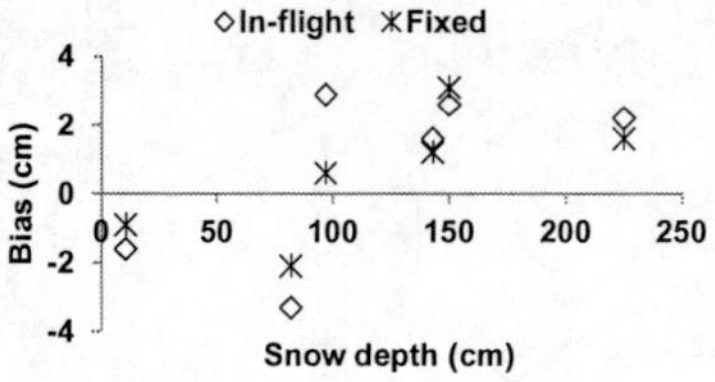

Figure 7. SD bias values for good light conditions.

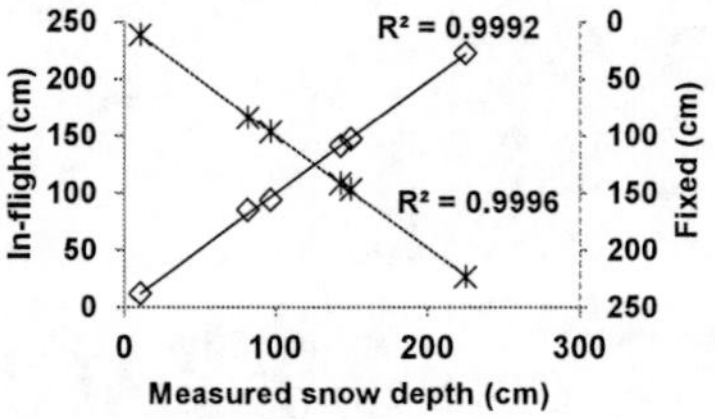

Figure 8. Scatter diagram of SDs.

Data availability. Data are currently not publicly available due to privacy.

Author contributions. AET: Conceptualization, Methodology, Visualization, Writing – Original Draft, Writing – Review & Editing SD: Funding acquisition, Investigation, Project administration, Writing – Original Draft, Writing – Review & Editing.

Competing interests. The authors declare that they have no conflict of interest.

Special issue statement. This article is part of the special issue "Earth Observation for Integrated Water and Basin Management: New possibilities and challenges for adaptation to a changing environment". It is a result of The Remote Sensing & Hydrology Symposium, Cordoba, Spain, 8–10 May 2018.

Acknowledgements. This research was supported by Çankırı Karatekin University BAP project MF-200217B15. The ECMWF numerical weather forecast outputs of total cloud cover were a courtesy of Turkish State Meteorological Services.

Edited by: Michael Cosh
Reviewed by: three anonymous referees

References

Avanzi, F., Bianchi, A., Cina, A., De Michele, C., Maschio, P., Pagliari, D., Passoni, D., Pinto, L., Piras, M., and Rossi, L.: Measuring the snowpack depth with Unmanned Aerial System photogrammetry: comparison with manual probing and a 3D laser scanning over a sample plot, The Cryosphere Discuss., https://doi.org/10.5194/tc-2017-57, 2017.

Barnett, T. P., Adam, J. C., and Lettenmaier, D. P.: Potential impacts of a warming climate on water availability in snow-dominated regions, Nature, 438, 303–309, 2005

Bühler, Y., Adams, M. S., Bösch, R., and Stoffel, A.: Mapping snow depth in alpine terrain with unmanned aerial systems (UASs): potential and limitations, The Cryosphere, 10, 1075–1088, https://doi.org/10.5194/tc-10-1075-2016, 2016.

Colomina, I. and Molina, P.: Unmanned aerial systems for photogrammetry and remote sensing: A review, ISPRS J. Photogramm., 92, 79–97, 2014

De Michele, C., Avanzi, F., Passoni, D., Barzaghi, R., Pinto, L., Dosso, P., Ghezzi, A., Gianatti, R., and Della Vedova, G.: Using a fixed-wing UAS to map snow depth distribution: an evaluation at peak accumulation, The Cryosphere, 10, 511–522, https://doi.org/10.5194/tc-10-511-2016, 2016.

Erxleben, J., Elder, K., and Davis, R. M.: Comparison of spatial interpolation methods for estimating snow distribution in the Colorado Rocky Mountains, Hydrol. Process., 16, 3627–3649, 2002.

Harder, P., Schirmer, M., Pomeroy, J., and Helgason, W.: Accuracy of snow depth estimation in mountain and prairie environments by an unmanned aerial vehicle, The Cryosphere, 10, 2559–2571, https://doi.org/10.5194/tc-10-2559-2016, 2016.

Manyoky, M., Theiler, P., Steudler, D., and Eisenbeiss, H.: Unmanned aerial vehicle in cadastral applications, Int. Arch. Photogramm. Remote Sens. Spatial Inf. Sci. XXXVIII-1/C22, 57-62, 2011

Nolan, M., Larsen, C., and Sturm, M.: Mapping snow depth from manned aircraft on landscape scales at centimeter resolution using structure-from-motion photogrammetry, The Cryosphere, 9, 1445–1463, https://doi.org/10.5194/tc-9-1445-2015, 2015.

Papa, F., Legresy, B., Mognard, N. M., Josberger, E. G., and Remy, F.: Estimating terrestrial snow depth with the Topex-Poseidon altimeter and radiometer, IEEE T. Geosci. Remote Sens., 40, 2162–2169, https://doi.org/10.1109/Tgrs.2002.802463, 2002.

Rinaudò, F., Chiabrando, F., Lingua, A., and Spanò, A. T.: Archaeological Site Monitoring: UAV Photogrammetry Can Be An Answer, in: International Archives Of The Photogrammetry, Remote Sensing And Spatial Information Sciences, Vol. XXXIX N. B5, 583-588, ISSN 1682,1750, 2012

Robinson, D. A., Dewey, K. F., and Heim, R. R.: Global snow cover monitoring: an update, B. Am. Meteorol. Soc., 74, 1689–1696, https://doi.org/10.1175/1520-0477(1993)074<1689:GSCMAU>2.0.CO;2, 1993.

Vander Jagt, V. B., Lucieer, A., Wallace, L., Turner, D., and Durand, M.: Snow Depth Retrieval with UAS Using Photogrammetric Techniques, Geosciences, 5, 264–285, https://doi.org/10.3390/geosciences5030264, 2015.

Westoby, M., Brasington, J., Glasser, N., Hambrey, M., and Reynolds, J.: Structure-from-Motion'photogrammetry: A low-cost, effective tool for geoscience applications, Geomorphology, 179, 300–314, 2012.

Zarco-Tejada, P. J., González-Dugo, V., and Berni, J. A. J.: Fluorescence, temperature and narrow-band indices acquired from a UAV platform for water stress detection using a micro-hyperspectral imager and a thermal camera, Remote Sens. Environ., 117, 322–337, 2012.

AUTHOR INDEX